TEXTING TOWARD UTOPIA

TEXTING TOWARD UTOPIA
Kids, Writing, and Resistance

BEN AGGER

Paradigm Publishers
Boulder • London

Copyright © 2013 Paradigm Publishers

Published in the United States by Paradigm Publishers, 5589 Arapahoe Avenue, Boulder, CO 80303 USA.

Paradigm Publishers is the trade name of Birkenkamp & Company, LLC, Dean Birkenkamp, President and Publisher.

Library of Congress Cataloging-in-Publication Data

Agger, Ben.
    Texting toward Utopia : kids, writing, and resistance / Ben Agger.
    pages cm
    Includes bibliographical references and index.
    ISBN 978-1-61205-307-3 (hardcover : alk. paper)
    ISBN 978-1-61205-316-5 (Institutional eBook)
    1. Internet and youth. 2. Text messaging (Cell phone systems) 3. Computers and literacy. I. Title.
    HQ799.9.I58A35 2013
    004.67'80835—dc23
                                                            2013011453

Printed and bound in the United States of America on acid-free paper that meets the standards of the American National Standard for Permanence of Paper for Printed Library Materials.

17 16 15 14 13   1 2 3 4 5

Contents

Preface and Acknowledgments

Several years ago, when I began this book, the working title was *Blogging toward Utopia*. And then texting happened, and I updated my technological metaphor. That makes me wonder whether we will still have Facebook, or even the Internet, in ten years. I think we will have the Internet or some other global technology of electronic communication and information. Yet one must be cautious about reifying technologies that are here today but gone tomorrow.

Nevertheless, there is no going back; we cannot put the genie of pixelated communication back in the bottle. The Internet and smartphones separate the modern and postmodern, although we carry forward aspects of modernity such as capitalism and its culture industries. Much as I began to do in my book *Postponing the Postmodern*, I continue to reckon with the shifting boundaries between modernity and postmodernity—in this book, with particular reference to children and writing. One of my conclusions is that we are all children in the sense that we are all needy and not fully formed. Everyone requires a parent, which is to say that we need a solid and stable mooring in a world that doesn't change too quickly. And we need to be loved.

If all writing is autobiography, it is perhaps worth noting that I began to think about such issues when my kids were still in high school. I published an article presaging this book: "Text Messages: Reading Kids' Writing Politically" (2009b). Since I began writing about fast capitalism more than twenty years ago, I have been interested in the ways that discourses such as science conceal their literary nature—the fact that they have been authored—and do not simply reflect

VIII PREFACE AND ACKNOWLEDGMENTS

nature. Concealment usually takes place under the guise of methodology, which is a cleansing exercise. I call writing that conceals its literariness "secret writing." Kids' writing, such as texting and Facebook posting, confesses its literary character; if anything, it overshares. Oversharing is interesting in its own right; it reflects the fact that people are alone and seek connection, even if they don't always choose the best means to connect. Almost like an anthropologist from Mars, I have observed my scientific colleagues write this way (and then conceal it). All secret writing can be read as the authored act it was. But to suggest that writing proceeds from literary perspective does not rob it of validity; all writing tells the truth, even when it falsifies or conceals.

I have learned from my kids and their friends that this is the most literary of ages and that they have a lot to say. But just as positivists don't consider science to be writing, many judgmental adults view texting and Facebook as too chatty to be writing. For them, writing lies in a narrow band of literary expression within which we can find Shakespeare and Robert Frost. The Martian anthropologist learns not to be condescending but to immerse himself in the world as he finds it. I hope that this book embodies respect for and empathy with my research subjects, the kids who check their phones before, during, and after my lectures but who can multitask, paying attention while checking over their shoulders electronically. We are not different; members of my generation were young once.

This is all late-breaking news. As I was preparing my final manuscript, the *Chronicle of Higher Education* (May 1, 2013) ran a story by Dan Berrett entitled "Students May Be Really Reading, But Not for Class." He reports on recent research by some Texas education faculty at Midwestern State University (SuHua Huang, Phillip Jeffrey Blacklock, and Matthew Capps) on college student reading habits. They find that 40 percent of what students read (sometimes during class time) is from social media, if not always from their college textbooks. My own experience is that students move back and forth fluently between Internet reading and pulp reading, documenting their term papers with citations from both media. It may be unnerving for pre-Internet faculty to observe their students poring over phones, tablets, and laptops, but this is literary activity and all literary activity is good—for the soul and for democracy.

This book responds to the fact that there is a great deal of psychic distress in the land. School shootings and youth suicide are but two manifestations of this. Many adults are miserable, too, and they either pass on their misery to their kids or fail to nurture them sufficiently because they are dealing with their own demons. Repeatedly, here as elsewhere, I call for slowing things down, working against the tide of fast capitalism. I call this "slowmodernity," signifying a blending of what is good about the modern, postmodern, and pre-modern. And I address the roles that people can play in achieving that slowmodernity, including how we write and form community.

I am enough of a Marxist to believe that the manifestations of psychic distress, as well as the loss of face-to-face community, are structural outcomes. To fix particular things such as schooling and children's diets, we must fix the totality. Band-aids don't work. And yet we cannot change structures without simultaneously changing the particular ways we live our lives in the present. People make choices, and can make better ones, even though "choice" is constrained by structure. Films such as *Forks over Knives* demonstrate how switching from meat-based to plant-based protein can change the world, one life at a time. Changing childhood and schooling can likewise have an enormous impact on overall social and economic structures, in effect creating "new sensibilities," as Marcuse called them in his 1969 manifesto *An Essay on Liberation*. And these new selves can and do form and join social movements such as Occupy, provoking certain authoritarian responses from those who have a lot to lose when "youth [are] in revolt" (Giroux 2012).

This book is anchored in my experience of the 1960s, a theme I took up in my 2009 book *The Sixties at 40*. Some people prefer vertical relationships, with sycophancy and subordination, and some prefer horizontal relationships, where everyone has value and is valued and people don't stand on ceremony. I'm a sixties person and always will be. I think we should make society less hierarchical and train people to be democratic citizens who enjoy the type of egalitarian relationships I am talking about. A great place to begin this retraining is by valuing kids and by listening to them, taking them, their writing, their angst, and their imagination seriously. I hope that *Texting toward Utopia* is a step in that direction.

*   *   *

Dean Birkenkamp, as he does, waited patiently for this overdue book. In its pages, I talk about various models of publishing, academic life, and schooling, including as sausage factories! Paradigm Publishers is anything but a sausage factory; the house is run by and for intellectuals. In these pages, I worry about the decline of books and publishing. Dean and his team indefatigably resist this trend.

Editing is a crucial part of a vital public intellectual life, not only for the obvious reason that editors edit authors. Editing teaches one to be cosmopolitan, to have broad tastes and to wander off the beaten path, outside of one's own language game. I have spent much of the past twenty years editing others' work as well as writing my own. My work with the authors who write for *Fast Capitalism*, which I coedit with Tim Luke, has been enormously enriching and has helped me broaden myself.

Several people at various stages of the intellectual gestation of these ideas provided invaluable support and useful feedback. Norman Denzin, Robert Hassan, Jeremy Hunsinger, Tim Luke, and Mark Worrell deserve thanks.

Finally, my kids, Sarah and Oliver, like my many students at UT-Arlington, have shown me how to be generationally cosmopolitan and, I hope, compassionate. As this book reveals, I am quite concerned that parents and educators are doing significant harm, even where they may have good intentions.

PART I

NEW SITES OF
WRITING, WRITERS, WRITINGS

CHAPTER 1
THE BOOK UNBOUND

The printing press helped end the Middle Ages, which restricted literacy. The Internet is on the verge of ending books, as we have come to know them. By "book" I mean considered reflection on the world that is produced as a readable object. To be sure, computer downloads may count as books, even if they remain only on the screen. But books, to earn that name, must be considered slowly and at a certain distance from everyday life. They have a spine, which holds them and their arguments together. For a book to have a spine promises distance from the everyday world required to consider its writing carefully and to formulate a rejoinder, the essence of democracy and community. Even these words may be anachronistic as fast capitalism has recently brought us books on the screen.

The decline of books is paralleled and reinforced by the ascendance of the mobile phone, which combines the functions of the laptop computer and traditional telephone. BlackBerries, Droids, and iPhones allow for, indeed, seem to compel, compulsive connectivity, combining talking, texting, tweeting, blogging, and surfing in a portable unit even smaller than a paperback book. People author their lives using phones, which allow typing, but this writing for the most part immerses them in everyday life and does not encourage distance from it. Users do not compose; they chat and disclose, spewing forth what Adorno (1973b) called the "jargon of authenticity" in his early critique of subject-centered philosophies such as existentialism. The unbinding of texts, replaced by multitasking phones, represents the triumph of connection over reflection, perhaps a natural outcome of postmodern alienation.

The shift from reading bound books while sitting or slouching to reading books on the computer screen—or not reading at all—constitutes an important aspect of the shift from modern to postmodern eras, from Fordism (mass production) to post-Fordism (flexible, just-in-time production), from reason to its eclipse. I want to avoid condemnation here; one can stare at the screen, even at the risk of postural problems, and treat pixelated argument in the same way one considers pulp. I'm doing that very thing as I compose these words! But in staring at the screen one is tethered to the technology. And one loses the sense of the book as a totality of sense and sentience—held in one's hand, thumbed through, dog-eared, annotated, read and re-read.

That is not the only problem involved in the unbinding of books. Not only does reading change, but writing changes. Given the attentional and postural challenges of reading onscreen, whether the smartphone, tablet, laptop, or desktop, writing simplifies itself, both in form and content. Text messaging and tweeting are examples here, as keystrokes are restricted. Try composing a text containing the word "epistemology" more than once! Younger writers resort to emoticons and the quickspeak of acronyms in order to compress their arguments. This is strange because a literary political economy would seem to promise almost unlimited text in an electronic public sphere. At issue is not just the restriction of keystrokes but the attenuation of attention in a posttextual age.

And yet. This is the most literary of ages, as I explore in this book. Writers, especially youngsters, write furiously, in response to alienation and toward utopia. It is not unrealistic that a young person would compose one hundred text messages a day. Assuming that each text contains an average of ten words, this amounts to one thousand words per day. In only seventy-five days, the texting author—for that is what she is—would write the equivalent of a 170-page printed book. Books of a sort are being written, but they take nontraditional forms easily ignored or dismissed by pre-Internet adults, even—no especially—by the kids' own parents. This book explores these changes in the ways we connect and communicate. My main thesis is that young people don't write or even read traditional books but they still express themselves in literary ways. And I'm not talking only of the young; boundaries between generations blur as adults do many of the same things. Perhaps the most relevant boundary is between pre-Internet and Internet generations.

The history of the book has been well discussed and continues to be of interest as we enter a posttextual age. Of particular concern has been the impact of printing, publishing, and librarianship on readers and writers, a central feature of modernization as we know it. Scholars, including historians, students of library science, and even social and cultural theorists have written on these issues in books and journals like *Book History*. Dahl (1958) offers a history of books, while later, more theoretically inflected treatments, such as Hall's (1996), examine the book as a vital component of culture.

The advent of the Internet seems to change everything—or does it? Turkle (1995) examines identity as people acquire their worlds and meanings from the computer screen, while Luke (1989) and Poster (2001) examine power and domination as these are increasingly pixelated. I have written about literary political economies, tying writing and other cultural issues to critical social theory (Agger 1990). The Frankfurt School first opened these questions when they (e.g., Horkheimer and Adorno 1972) introduced the concept of the culture industry, a Marxist opening to what later came to be called cultural studies.

The rise and decline of the book contain a fateful dynamic. As global cultural dissemination has been attained in our post-Fordist moment, we can get out the message but we have lost the distance of books from the realities they describe and discuss. Overcoming physical distance seems to have reduced critical distance required to appraise the world rigorously. It is fundamentally different to read a Wikipedia entry than an old-fashioned encyclopedia entry, an electronic book rather than the real thing, email or text messages rather than letters from yesteryear. Near-instantaneity has reduced the time it takes to compose and then interpret writing. This fateful foreshortening tracks the rise and decline of the book, which originally liberated Europe from myth and misery. Perhaps it is enough to say that the Enlightenment has gone too far or, better, that it was diverted—the original argument made by Horkheimer and Adorno in *Dialectic of Enlightenment*.

My analysis is awash in nostalgia, even for my own earlier literary career, when I read and wrote books that inserted themselves in the ongoing conversations about social and cultural theory. I still read and write, but one has the nagging sense that we are writing for a very few. Doug Kellner asked me recently whether I thought the Internet was the way to go as far as publishing, of books and journals, was concerned. I responded that many more people read me and him electronically than in pulp—that it is difficult to stay invisible on the Internet. Part of me values the Internet as nearly frictionless and accessible, a vehicle of cyberdemocracy. But the Nietzchean/Adornoian worrywart side of me frets that this will only deepen conformist thought, an ability to rise out of the ooze and think the world otherwise. It is easy to conceptualize the Internet as a surrounding, deboundarying ether in which critique harmlessly gets absorbed or alternatively to view the Internet as edgy and indie, a perfect vehicle for the long march through the institutions, as the German New Leftist Rudi Dutschke termed it. My doomsaying temperament is probably justified, given the trajectories of capitalism since the nineteenth century. An iPod/laptop capitalism is "totally administering," to borrow the Frankfurt School phrase. Everything is controlled and dissenters are disciplined. Yet weirdoes— "difference," in Derrida's terms—slip through the cracks and even occasionally flourish. Non-one-dimensional thought abounds here and there—sometimes in Europe, occasionally in Eugene, Ann Arbor, or Austin.

But the exceptions prove the rule: capitalism unleashes "domination" in order to keep people in line politically and in line at the malls—now, both in line literally and also online, a post-Fordist vehicle of commodity consumption. Domination is the heavy weight of conventional, consensual ideas that blocks the imagination of a different, better world.

Even before the Internet became an ether, I wrote along these lines in my book *Fast Capitalism* (1989a), which tracked the decline of discourse as the end of the book. In a sense, things have gotten worse, but they may also have gotten slightly better—my answer to Kellner, who probably shares my ambivalence. Indeed, we both dislike aspects of the traditional pulp/publishing world, already named the culture industry by Horkheimer and Adorno. Part of one-dimensionality is banality, but banality driven by the relentless logic of the market, which both reflects and reproduces an affirmative culture. Although that might sound pejorative and even elitist (a consistent critique of Adorno's aesthetic theory), by "affirmative culture" I am using a technical term to describe the reduction of thought and hence culture to clichés, tropes, simple sentences—exactly what we observe as we track the decline of a public intellectual life (Jacoby 1987). Who were the first affirmative thinkers, publishers, writers, and readers? In a sense, it is does not matter; these agents are arrayed dialectically. Publishers claim that the market (readership) makes them publish banal, uncritical works. Authors contend that their challenging prose has been domesticated needlessly. Curmudgeonly readers lament that they can find very little worth reading.

The rise of onscreen reading and writing has led to some important changes in reading culture:

- People read via the Internet, downloading information and entertainment.
- Bookstores are in decline, and independent booksellers and publishers are failing.
- People write—blogging, texting, messaging, posting—but many epistles and screeds float off into cyberspace, not matched by accompanying readings.
- Writing in pulp formats is increasingly formulaic and scripted, parroting the prevailing norms of the market.
- In academia, people write in order to get published, not to be read. Technical language abounds.
- The decline of public intellectuals is matched, and hastened, by the decline of public readers curious about the state of the world and passionate about changing it.
- The Internet affords access and enhances accessibility, where it is not commodified.
- The posttextual replaces dense and closely argued prose with images that summarize, simplify, and stimulate. Much Internet traffic involves the imaging of bodies and sexuality.
- Publishers, both trade and academic, feel compelled to publish what will sell. Loss leaders and niche books lose their footing.

Marketing replaces editorial development as an activity of publishers and journalists who feel the pinch of the Internet and print-on-demand. These transitions cause the rate of intelligence to fall and discourse to decline (Jacoby 1976). This is not to draw a firm boundary around the post-Gutenberg, pre-Internet era, when books prevailed. There was always the tendency, within that era, for culture industries to commodify writing and render writing banal. And the spread of the Internet, since the late 1980s, did not instantly cause publishing houses to shut down and library budgets to shrink. These are boundary crossings, tendencies. Diligent scribes still compose for pulp publication and use the Internet for the dissemination of considered writing, writing at a "distance," in Adorno's terms.

But these transitions represent powerful tendencies for "the book" to come unbound, for publishing to become entertainment, and for the very acts of writing to change—composition becoming tweeting, posting, texting. Even the blog, perhaps the most traditional postmodern form of literary craftsmanship, is designed more to be written than read. After all, few care about your cat in Topeka or your tumultuous dating life or your views of Obama and Palin.

Celebrants of the Internet (e.g., Negroponte 1996) prophesy a digital democracy. Connectivity could expand the New England town meeting to a global polity. It could also break through the walls of the local library and even the Library of Congress. Everything would be available, and every voice could be heard. And the opportunity to blog, text, and tweet makes each of us an author.

Although these tendencies exist—I coedit an electronic journal and rely on the Internet for communication and as a research tool—there are powerful countertendencies, such as commodification (the tendency to put a price on everything) and conformity, identified by Marx and the Frankfurt School as the tendencies of what Marx termed the "logic of capital." Lukacs and the Frankfurt School amplified Marx's nineteenth-century argument in explaining why the socialist revolution that he reasonably expected got sidetracked. Their answers lay within Marx himself, notably his argument about "false consciousness"—a systematic belief system that fundamentally misrepresents the world and foreshortens the person's freedom, unnecessarily.

First Lukacs (1971) in his concept of reification and then the Frankfurt School in their writings about domination and the culture industry argued that false consciousness has been deepened, especially in a post–World War II consumer culture. The sale of commodities necessary for survival is not sufficient to sustain capitalism. Now, capitalism must inculcate "false needs," encouraging people to spend beyond their means using credit on indulgences and entertainments (Marcuse 1964). These false needs are redefined as necessary, both because one must keep up with the neighbors and because technological prostheses (think of television, the Internet, cell phones, automobiles) are portrayed as inevitable concomitants of modernity—what people must have and use in order to be modern or, perhaps, postmodern.

One might define postmodernity as the eclipsing of books, basic needs, Fordist factories, and boundaried nation-states. The postmodern can be celebrated as globalization, in which the Internet plays a major role, but a Marxist notices that globality is simply the continuation of class struggle by other means. Marx and Lenin already understood international imperialism and colonialism as essential for European and American capitalism. The outsourcing of jobs, commodities, and culture to the Third World perpetuates uneven development, on which capitalism rests.

What is genuinely different about this scenario from when Marx and Lenin were writing is that countries such as China combine economic development with political authoritarianism. Marx thought that the advent of the industrial age would bring democracy, although a spurious representative kind that would collapse under the weight of inevitable economic crisis and lead to real democracy of the communes and the soviets. China and Russia demonstrate that Marx and Lenin's developmental scenarios were not exhaustive of historical possibilities. These countries combine economic development—consumer capitalism, the Internet, culture industries—with political illiberalism, suggesting that there are alternative models of modernist development, some of which might be termed postmodern. China and Russia might be "post" in the sense that they outlive Marx's and the Frankfurt School's essentially Hegelian optimism about world-historical Reason as materialized in communism. Habermas extends this utopian tradition by urging the completion of the project of modernity, not its abandonment. But capitalist connectivity is not necessarily accompanied by democracy, justice, and a universal regime of Reason.

Indeed, what we are seeing, and not only in modernist/authoritarian regimes but also in the parliamentary West, is an admixture of consumer and entertainment capitalism, based on highly portable connectivity, and massive depoliticization and anti-intellectualism. People chatter and stay connected but about ephemera—precisely the concern of the Frankfurt School in their culture-industry and one-dimensionality arguments, and as amplified by Jacoby, myself, and others who discuss the decline of discourse in a fast, perhaps arguably postmodern capitalism. Habermas (1989) addresses these issues as he discusses the structural transformations of the public sphere in late capitalism.

The tea leaves are difficult to read. The thesis of the eclipse of reason founders on the evidence that this is among the most literary of ages, at least if one simply tallies keystrokes per day per person. People of all generations, such as the young using Facebook and Twitter, produce thousands of words as they get and stay connected. Are these words ideas? There is no reason they cannot be. They don't usually achieve distance in the quickspeak and code of instant messaging and texting. Adorno would not have sanctioned "ticket thinking" such as LMAO or LOL. He wouldn't have endorsed the use of emoticons. Perhaps this is a stodgy point of view in today's fast world, in which the text message replaces the paragraph.

Both things could be true at once: there is a monumental and global dumbing down, but there is also frenetic literary activity as people write—both their "selves" and in connecting to others. How ought we to read the compulsion to write and reach out? A technological determinist might simply note that the technology is there to be used, and we use it, much as supposedly labor-saving vacuum cleaners actually increased women's labor after World War II. But I think there is something deeper, especially among the young. This busy writing constitutes protest and a yearning for community, a tapping on the walls of their cells as young people create a world below the adult radar screen, both in protest and in the building of community.

To use the pre-post-Fordist Marxist language, these busy scribes—bloggers, texters, tweeters, posters—are alienated and they are responding by writing their alienation. They communicate in code because they don't want parents and their teachers to have access. These are the language games of rebellion, even if Marx and Adorno could have scarcely imagined a proletariat comprising generations X, Y, and Z who constitute a pre-labor force, kept busy by a long school day, homework, and "activities" positioning them to succeed in the adult credentialed world.

The adolescent "lumpenproletariat" (Agger and Shelton 2007) is matched by alienated adults who spend much of their waking time online. A postmodern deboundarying also affects the thinning boundary between work and home/family/leisure. Phones that double as computers allow the sort of fast literary craftsmanship I am talking about. Adults sit side by side in waiting rooms working with their computers or phones. Paid work and unpaid activities bleed into each other as people open multiple windows and bounce back and forth between what, in an earlier modernity, were physically and temporally separate spheres.

I purchased my first cell phone—a $20 pay-as-you-go phone that I fill with purchased minutes. My kids urged me into postmodernity, and I taught a course on fast capitalism when I made the buy. I told my children and my students that I'd give it a month in order to see whether my life changed in significant ways! Perhaps predictably, the phone is already an alienation: I have to keep track of it, and it compels me to answer it and to check messages. It creates work and sucks up time, even as, one must concede, there are certain efficiencies and utilities, such as keeping track of my kids and communicating with my wife. But I waited until I was fifty-six to do this. I remained pre-postmodern, and I don't think I was missing out on much.

Americans are said to watch four hours of television a day. I suspect that this number will decline, now that people can, in effect, write their lives using rapid information and communication technologies. These tools consume time, perhaps borrowed from paid work, television, parenting, sleep. Books were never this compelling, except when we found a good read that we couldn't put down. We could always dog-ear the page and come back to it. Indeed, not reading straight through heightened our anticipation of plot development and denouement.

Adorno wanted writing to be dialectical, mirroring the contradictions of the world. Music (Adorno 1973c) of a certain kind (for him, Arnold Schoenberg's compositions) could do the same thing, allowing us to approach truth by remaining distant. His own sentences were models of allusion and indirection. One has to work at them in order to understand the ways in which they track the world. But few read Adorno or listen to Schoenberg. Even popular books risk becoming museum pieces—what existed before the Kindle.

The unbinding of books is itself a dialectical phenomenon. It cheapens the production of books, and yet it also attenuates writing and attention. Literary life on the screen is thin, even one-dimensional, unless we download, staple, and perhaps bind. And even if we do that, we are assuming that writing remains distant, not sausaged into a few hundred keystrokes and littered with computer code and emoticons. Literary life is impoverished by comparison with writing before the Internet, even if publication—in the broad sense of getting your wares out there—is less expensive.

Must books have spines? Must authors have spines? A tentative yes to the first question and an emphatic yes to the second. The global, instant technologies of cultural production and transmission need to be set in the historical contexts of the pre- and post-Gutenberg worlds. Setting type changed the world, and now hitting "send" and "save" may have an even greater impact.

In the following chapter, I consider kids' "secret writing," the ways they compose themselves in the post-Gutenberg world. Although the unbinding of books has distressing elements, all is not lost: Kids write furiously, as they protest, connect, and even imagine and work toward a better world.

Chapter 2
Secret Writing

When college students leave class, they check their cells, searching for calls missed or messages left. Some initiate contact or respond to messages. My teenage kids are always keyboarding and texting, staying in touch with their worlds using instant messaging, Facebook, even email. Young people describe their comings and goings using Twitter—producing a descriptive ethnography of their everyday lives. They are connected in ways I never was, although I could walk down the street and find a friend in an unlocked house. Times have changed.

Kids Write and Read Digitally

The young and, increasingly, elders use phones as computers and cameras. iPhonified discourse blends and blurs writing, talking, filming. On the surface, these are devices of convenience, freeing people from their desks and enabling multimedia presentations. They also enable nearly instantaneous connections. Underneath the surface, though, it is not clear that this is progress, especially where people, especially the young, lose touch with considered "pulp" (paper-based) writing and trade cybercommunity for the face-to-face kind. At stake are democracy and the literary self.

What is democracy? What is writing? These questions animate this book. In general, my perspective is that democracy is dialogue and that writing is the considered opening of the literary self to others—readers, who would become writers. There is much to appreciate about screen capabilities. They empower young people to become young authors—not only of texts but of their own lives. Kids' writing needs to be valued by adults for whom text messages, email, and blogs are necessary evils that suck time away from "real" work. Part of the challenge posed by our postmodern moment is to trace the boundary between the real and unreal or surreal, perhaps deciding that boundaries need to be eased and sometimes pierced altogether. Writing oozes out of its covers and into the world itself, as mobile discourse demonstrates. To be "old" (in the eyes of young people) is perhaps to insist on boundaries, almost for their own sake.

We responsible parents and teachers insist that there must be boundaries lest postmodern young people ooze out into the world, dispersed and decentered, much as their writing is. I am not championing "deboundarying," but as certain boundaries dissolve, we run the risk of decentering selves, writing, and communities just as we gain the potential to reconfigure our lives, our texts, and our democracy. The Internet and these microtechnologies of power and practice cannot be put back into the bottle; they are here to stay. However, they can be theorized and hence mastered. This is particularly important for young people for whom certain types of deboundarying are taken for granted; they did not know our earlier worlds of stable selves, pulpy writing, and town meetings.

Deboundarying also provokes its opposite—the way in which kids erect new boundaries around their lives, especially to keep out adults, who are frequently experienced as aliens and agents of alienation. Perhaps the interesting question is not whether there should be boundaries (such as between the public and private) but *whose* boundaries, and *for what*. For much of the past, boundaries kept people out, as well as enclosed those on the inside. Here, the deboundarying of writing, its oozing beyond the writing desk to anytime/anywhere pixelation, opens literary opportunities to groups heretofore denied the opportunity to author texts and their own lives—the young. This is not to valorize all the writing that kids do, but to value and nurture the opportunity to write, which is what texting and Facebook afford the young.

These opportunities are opened to elders as well. As of this writing, just over half of Facebook users are between ages eighteen and twenty-five. Use is growing rapidly among older people. People born before 1960 will be slower to come to these forums and they may never arrive there. I don't Facebook, although, having kids and students, texting has been thrust upon me as having a certain economy, if not inevitability. My wife, born at the height of the baby boom and just before 1960, has started to Facebook, using it mainly to share photos with family and friends. At a time when families have been dispersed (a kind of deboundarying of families and local communities), social-networking technologies offer a new type of community, requiring us to think in new ways about families and virtual selves (Agger 2004b).

Have these technologies affected student learning? Many students dislike school, because teachers pile on homework and busywork, heedless of the consequences such as sleep deprivation and the attenuation of childhood. Not all schools or teachers do this, but this is one way to explain the anti-intellectualism of our culture at large. We teach kids implicitly that questions have only one answer, which can be memorized. And writing is formulaic, to be completed using "writing trees." Many educators, even at the college level, teach to the test. With the Internet, students gain access to the global world and write thousands of words a day. And they gain new skills, such as how to manipulate computers, surf the web, and design web pages. But too many don't know anything about the war in Vietnam or the civil rights movement or who Stalin was. The world for them has neither history nor mystery. My own college class attendance is far from perfect. I joke about this and teach to those who are present, but it bothers me. Why have kids lost the love of learning? My students profess to like college more than high school because, they say, they are free and many faculty respect them, unlike in high school, which they experienced as a prison.

This is a story in which there are no villains. Or perhaps everyone bears responsibility. Adults—teachers, parents, school administrators, cultural critics—bemoan children's lack of literacy and their minuscule cultural capital. Kids hate school, homework, writing and reading, adult authority. Adults ratchet up the homework and testing; they want school uniforms and a uniform curriculum. Kids rebel by texting each other in school—inmates tapping out code through the walls of their cells. A double entendre: cell phones are imprisoning, requiring constant attention and distracting kids (and adults) from the task at hand, including driving. It seems that everyone is chattering and complaining, but that no one is listening or theorizing. I am pro-kid, but I am also a cultural critic. I love kids, including my own, but I worry that my college students don't know anything. Well, they know things I don't, like how to search online effectively and build websites; but few know what caused World War II or even who the sides were. They are long on posttextual acumen, but short on textual attention span. They don't want to read books without images; indeed, images have become texts. They consult Wikipedia and they prefer the thin knowledge of the Internet over the thick knowledge of books. They type and text quickly, and they clearly love writing, even if the writing they do isn't writing by our standards—with old-fashioned punctuation and grammar.

Bound books may be a thing of the past, but nevertheless they are the past; they represent an indispensable cultural and intellectual history. One cannot be civilized and educated without reading through the vast traditions that make the past present. But it may also be true that the book, if not entirely a dinosaur, is not the only available form of literary self-expression. Our kids use other modes and their writing is important and needs to be read and appreciated, even if it often defies our own literary standards and cultural preferences. Secret writing is writing that does not fit the mold of writing since the Greeks and of publishing since Gutenberg. Such writing flies below adult radar. Generations blur but are still distinct as adults learn to text and use Facebook, reading and writing like their children but still evaluating kids in terms of traditional standards of literacy.

Not everyone ignores kids' writing. Gloria Jacobs (2009, 2008a, 2008b) studies kids' writing in a serious way. But for the most part teachers and scholars decry texting, instant messaging, posting, and phoning as inauthentic modes of literary self-expression that actually divert kids from "real" literature and authentic literary activities such as reading and writing. Part of me favors that critique, but, as a parent, part of me realizes that we must confront the facts: kids write furiously, and yet we dismiss or ignore this writing. Scholarship has tended to ignore kids' writing because it ignores kids. It could be said that children are an important but invisible minority group. That group is populous and global, and, like other minorities, its members are structurally disadvantaged: kids cannot vote; they lack legal independence; they are dependent; the worst adults treat them like slaves, prisoners, or chattel. Even when there is love, it is often mixed with paternalism, which perhaps undercuts genuine affection and empathy. Children are the last major but largely unrecognized minority group. This is ironic because we were young once, and we forged the sixties, which was a children's movement. How quickly we forget the young and their efficacy.

The Concept of Childhood in the Digital Age

Until the Industrial Revolution and Victorianism in the nineteenth century, children were treated and viewed as small adults. They worked alongside parents in the home and on the farm, and their chances of surviving birth were not impressive, given the lack of prenatal and neonatal care. With Victorianism, which induced women and children to leave the factories and retreat to the domestic sphere, childhood was framed as a distinctive developmental category, and suddenly children were precious and dependent. They received Christmas presents and birthday cards. Women became "moms" responsible for managing the distinctive phase of childhood development.

This boundarying of childhood held until the mid- to late twentieth century, which inaugurated laptop capitalism. As women entered the paid workforce and the public sphere in significant numbers, home life lost some of its centrality. Women who worked for wages outside the home were expected to perform what Hochschild (1989) calls the "second shift" when they returned home: caring for children, cleaning the house, making meals. Life became hectic, and without a corresponding uptake in domestic labor by men, many children increasingly had to fend for themselves. Latchkey kids return from school to empty houses.

Other trends have contributed to children's forced self-sufficiency. As both adult work hours and children's work (homework, activities, lessons) have expanded, the family home has become a frenetic work site. The workplace replaces the family as a haven in a heartless world. Divorces have made single-parent households common. Children are saddled with many adult expectations about their performance, but they are often left to their own devices as they try to meet these expectations. They are sleep-deprived and they search for electronic narcotics such as video games and television. Too often, children are left alone by adults who are living their own busy lives.

These electronic connections among the prepubescent and pubescent proletariat should be read not as Morse code but as literary work. It is potentially political work—resistance and organizing—as children share their grievances and form countercommunities. High school students place their phones in the little pouch of their hoodies, allowing them to disguise their busy manipulation of the text keys at school. Texting is code and easily concealed from adults. Schools confiscate phones if these are seen by teachers and administrators. After Columbine, having cells during the school day would seem like a necessary expedient, indeed a means of connection and comfort for parents and children worried about school violence. But the schools have come to understand that kids are forming countercommunities and engaging in resistance as they use their phones to gossip, complain about adult authority, and make social plans.

Parents often beseech their children to use their cells for talking, not texting, in order to save money. One text message leads to many others, as the number of keystrokes allowed limits old-fashioned discursive paragraphs. Texting involves a peculiar literary economy of limited keystrokes and economizing codes and acronyms: LOL, BRB. Parents urge their children to talk, not text, in the interest of efficiency. But kids choose the more time-consuming literary work of texting and tweeting. They prefer the concealment afforded by texting, which allows them to mediate their communication in ways unavailable if they talk, whether voice to voice or face to face. The countercommunity of children and adolescents has aspects of adult community, but it differs in that children, perhaps in their nature, are more comfortable with concealment and mediation, offering less of themselves to others.

Perhaps there is less "self" to offer. The teenage years are spent in pursuit of authentic identity, a destination only available to those who try on multiple identities, who experiment with who they are. Electronic communication is perfect for these transformations and trials. Screen names screen the self and allow kids to try on identities. Adults do many of the same things, as we can see in studies of Internet dating (Vitzthum 2007; Agger 2012). Dating sites have their own codes, like "height-weight proportionate" for "not skinny but not obese." People spin themselves as they become their onscreen profiles.

The decompartmentalization of adulthood and childhood, which reaches back to agrarian feudal life, has been accelerated in fast capitalism. Kids grow up quickly, and they are connected to, and burdened by, the world in many of the same ways as their parents. The Victorians had the right idea, if the wrong solution: children needed not to work in factories but to grow up playfully and in their own time. Turning women workers into moms, instead of sharing parenting between the sexes, proved to be a problem in the long run as women "liberated" by the women's movement and economic necessity to work outside the home remove themselves from aspects of parenting that were never done by men. A frenzied capitalism in which there are no safe havens isn't good for anyone, including parents who are themselves needy, especially after divorce. Adults also text, instant message, send email, use social networking, conceal themselves, try on new identities, lie. Adults also form countercommunities and speak in code. Just as children have become adult-like, adults have become needy in child-like ways. No one is getting their needs met.

Much of the story of childhood today is about time and time use. Ours is an age of distraction; distraction is a utopian impulse when the workload becomes too heavy. Kids perform adult-length work days in school and then return home for lessons and homework. They resist by straying off task and forming their own

cybercommunities. They are protesting and organizing. They are compartmen-
talizing what has been decompartmentalized. They want boundaries between
themselves and adults. They want boundaries between work time and free time.
Nighttime becomes their time, off the clock.

Why don't adults get these things? We were young once. When we were
young, we didn't have these communication technologies, and we didn't have
long work days and staggering amounts of homework. We wandered outside to
play. We formed community in the streets of our neighborhoods, in parks, and
on the playground. We worried less about college prep and resume building.
Sports was a choice, not an obligation. I never took a practice SAT test. My
parents rarely checked my grades. Playing sports was my choice, not theirs. We
were young once, and we remained young, until it was time to move on. We got
enough sleep and physical activity. Fast food didn't beckon. Most didn't have
cars and had never heard of porn.

The 1950s and 1960s were a golden age for many American kids, if less a
golden age for women and other minorities. Idealizing the past risks repeating it.
But the ample boundaries were healthy for children and helped them form stable
identities. In postmodernity, boundaries disappear as kids become adult-like and
adults, especially when they lose their partner, become needy and neglectful.
Or they simply work too much. This story is not only about time, though. It is
also about technology and capitalism and the ways that supposedly labor-saving
devices actually create more work and have unforeseen social consequences.
Digital technologies, abetting globalization by allowing people to work and shop
anytime/anywhere have certain advantages and economies, but they also remove
boundaries, especially between stages of childhood and adolescent development,
that are necessary for the growth of an autonomous self. We are all positioned
by the rapid information and communication technologies sold to us by companies that do not theorize their own responsibilities for the social consequences
of deboundaried, accelerated, anomic lives. Smartphones, laptops, and tablets
are the only games in town. My own university, to economize, removed some
faculty telephones, requiring students to communicate with faculty using email.

Nonetheless, kids are not only victims; they use technology to resist the
pressures on their own lives. Sociologists have neglected much of this because
children are not treated or viewed as full citizens and because we tend to view
childhood ahistorically, in biological/developmental terms and not in social/
political terms. Adult social scientists may also ignore these things, even if they
have children of their own, because they lament what they view as generational

decline, especially in comparison with their own golden childhood that may have opened into political activism during the sixties. Above all, we ignore the fact that children are agents with free will and efficacy. They are prisoners of a sort, but they resist and rebel. We use the term "slackers" for kids who are approaching adulthood but who eschew adult roles and performance expectations. Slackers are kids who reject adulthood and adult roles, even as they enter biologically and developmentally into apparent adulthood; they have adult bodies and urges and even economic means, but they don't "buy into"—a telling double entendre—the adult rat race. They are often derided as irresponsible. Older adults may feel that young adults are lazy and intellectually incurious, don't care about the world, and are making poor choices that will disserve them later on. We pick up on their resentment of us as we pile on the reading, writing, and other homework in an attempt to rectify them. And we detect their resentment of our silent judgment that we are superior to them, and were superior to them even in our adolescence and college years, when we were engaged politically and socially.

But there is another way to appreciate the slacker mentality. Slackers' lifestyles and choices represent a profound type of resistance that foments in the countercommunities and rebellion of adolescents who resist being dragged into alienated adult life. They recognize that their parents are not happy and may even be regressing to childhood. While they may have abandoned traditional forms of literacy and be late with their term papers, they have also spent the semester pouring out thousands of keyboarded and touchscreened words and read as much in turn. They surf the web, post on it, text, chat. They create web pages and they blog—a huge and invisible cottage industry that is transparently literary work. In my own classes I try to bridge the generations by assigning a journal. This is essentially a blog, where they record a couple of pages of their reactions to the readings and lectures every week. I secretly slip in some structure as I ask for a title—in effect, a mini term paper. They are encouraged to discuss the implications of grand social theory for their own lives, perhaps jobs they have had or relationships they have experienced. Many students eat it up, even as they resist writing traditional papers that require slow readings of pulp texts.

Children as Agents

The active, purposeful embrace of technology by young people is a strong counter to the view that kids are simply manipulated and dumbed down by digitality.

Young people rebel against their alienation, and against adults, by forming *countercommunities* and by producing *code* unfamiliar to adults. Young people have always forged their identity by rebelling against adult authority, thus achieving what Freud called individuation. But there are two differences between what is occurring today and generational rebellion in the past: adults are so preoccupied that kids are left to their own devices, affording time to form communities and develop code; and communication technologies are available to kids that allow them to escape adult surveillance.

Sometimes their rebellion resembles the anarchy of *Lord of the Flies* more than the sophisticated political and cultural subversion we might hope for them. In many cases, their activities are not so much transformation as escape, a resolution of personal anomie and not a social movement. The 1950s, though creating a safe haven for children in many ways, left them constrained by gender, clothing, and schooling rituals so suffocating that a whole generation rebelled during the 1960s. The social movements of that era—civil rights, antiwar movements, women's rights, and environmentalism—were children's crusades. It's not clear that the digital revolution is creating similarly transformative young people's movements today. Polls show that young voters played an important role in reelecting Obama in 2012, but often they simply avoid politics as a site of venality and cynicism, which is also the reaction of many adults after Watergate and Monicagate and Bush's eight-year-long folly. Being turned off politics, resenting authority, forming countercommunities, and developing code heard on a subversive, subterranean wavelength are the not the provinces of the young. The generations blur, just as they are separate. This simultaneous boundarying and deboundarying—children versus adults and children and adults occupying the same boat—require a nuanced dialectical analysis. I am not saying that kids are a new proletariat, a new and promising revolutionary subject, but that children are frequently as desperate as their parents and teachers for respite from routine, even for escape.

Marcuse (1964) in *One-Dimensional Man* urged a "great refusal" of late capitalism's disciplining, conformist tendencies. He particularly wanted people to reject needs fostered by advertising and responsive to people's deep-seated sense of dissatisfaction with their lives. Kids and many adults today are engaging in the great refusal, although they do not have the overarching theoretical framework recommended by Marcuse that would explain the source of their alienation and point them in the direction of a better society, to be achieved through activism as well as deliberate personal choices. In a word, the "refusers" today are not socialists, for the most part, nor feminists, nor other sorts of theoretical radicals.

Radicalism is in disrepute, as are all politics. Instead the slackers, much like the beats and hippies before them, are engaging in a pre-figurative, pre-political type of resistance and rebellion, auguring a better world in the countercommunities and code they created. A countercommunity is not (yet) a full-fledged social movement, and code is not (yet) theory. But they could become so, once they become self-conscious of their grounding in a longer tradition of radicalism and resistance. The teenager who texts his friends during class is far away from understanding that he is "refusing," prefiguring, auguring. But he could be shown that his personal troubles are indeed public issues, as Mills (1959) recommended in his manifesto for radical sociology *The Sociological Imagination.*

Kids think that they are unique, the first to bear their burdens, the only ones with paternalistic parents. Growing up is necessarily painful, as is what Freud called adult "repression"—inhibiting and then sublimating one's basic infantile impulses. However, children now constitute a class, or a pre-class, pitted structurally against adults, who treat them as members of a pre-labor force tasked with homework, lessons, activities—the assemblage of an industrious adult self. Without having a leader or a plan—except perhaps the musicians and music playing on their iPods—kids are marching to their own drummers; they slouch toward utopia, defined as the absence of adult expectations. This is not to embrace slacking as adequate political practice—indeed, in its own terms it can be self-destructive—but to notice that a pre-political pre-labor force of the young is engaged in an inchoate struggle.

Facebook is their town meeting, rap music their "Internationale." Their politics is an aversion to politics; their lack of commitment resembles that of James Dean. If they bothered to read Camus, they would recognize themselves as "étrangers"—outsiders. But when nearly everyone is outside, they are inside and slacking becomes the norm. The few who obey their parents and teachers and reproduce their productivism wind up at Ivy League universities, heavily indebted and heading rapidly toward adult alienation. If the slacking silent majority knew much about the 1960s, they would discover that hippie be-ins anticipate their own purposive purposelessness—chilling, as the young call it.

Chilling is the praxis of slackers who resist and rebel. They resist and rebel against performativity, the measurement and evaluation of all childhood activities. This is postmodern, uncoupling identity and work. It is Marxist, uncoupling worth and reward. It is existentialist, appreciating everydayness as a sufficiency of being. Adults, even progressive ones, worry that young people, and especially their own kids, will learn the hard way that those who lack ambition will slide backwards from their parents' social class. As if this matters.

The countercommunities and codes of the young refuse what I have called *performativity*—grading every behavior as if it were a test. And grading is inherently competitive in this zero-sum age; there are only so many As to go around. So-called model minorities such as Asians embrace performance because they come from pre-modern societies in which basic subsistence is at stake. Their kids are the last to text and post, unless they are totally Americanized, because such activities cut into homework time and SAT preparation.

The Literary as Political

This study seeks to understand young people's literary activities as political, as gestures of refusal and rejection. These are phenomena worthy of study by social scientists and of nurturance by progressive adults who view children and teenagers as a neglected minority group. The young are educable in the ways of political mobilization, especially if we teach them the lessons of the 1960s, when the young changed the world; we were the children then, and our kids are the children of the children. We need to listen to them, much as our parents should have listened to us during that turbulent decade. Let's turn up the volume and tune them in.

Seen another way, kids' writing, untraditional as it may be, is a message in a bottle, even a cry for help. Kids who text in school almost ask to get caught; they want adults to read their writing, to take them seriously, or at least to read the fact that they are writing subversively as proof that something is wrong: that the *Odyssey* turns them off, and what this means; that endless math problems are boring, and what this means; that they are being driven to distraction by complex schedule juggling and workplace expectations about copious and detailed homework and the testing done over "material" to be memorized and mastered.

Attention deficit disorder (ADD) medicalizes children's protest—the fact that kids are shutting down and shutting out the world. Drugs can help; there may be organic issues in play. But adults can help, too, by slowing down the pace, reducing the pressure, valuing creativity for its own sake. Childhood is being attenuated as we prepare kids for adulthood; this now begins in early elementary school, where there is homework, memorizing, testing, grading, workplace discipline. Kids want their world back: hence, their countercommunities without adults, and their codes that defy ready translation by adults. A friend of my daughter writes on her social-networking page that she is a "badass motherfucker." She doesn't mean that she commits crimes or violence; she is signifying that she is not who she appears to be, a good and docile student. She is restless and seeking another identity. It is insufficient to read these grandiose sentiments as typical childhood rebellion, mixed with self-glorification. These words are code; she is signaling to other kids that she isn't what she appears to be. She isn't only letting off steam; she is searching for community with other cynical nonconformists.

This seeking of community is frenetic literary work; it requires hours behind the laptop or desktop computer and on the cell phone. Electronically mediated dialogue, such as texting or messaging, is not for the faint of heart. It involves a serious investment of time, even to accomplish a single conversation. Kids return home from school and rush to their social-media sites, both to read and tell the latest, subjecting their lives to real-time scrutiny. We, from the sixties, although introspective, did not subject our everyday lives to such real-time scrutiny. We gossiped, but didn't post or tweet the gossip for all to hear or see. When we did, we used our parents' landline, standing in the middle of the house. There were few secrets, especially in small towns. There was no self-concealment, no aliases and screen names. Today, the Facebook generation lives in surreality—an almost inconceivable extent of sheer self-revelation in public cyperspaces. And they live in hyperreality—in the sense that the "reality" of social networking and messaging is perhaps more intense than "real" reality. We were not pure, but neither were we sur or hyper. Hyperreality hypes the real, exaggerating in order to capture its weirdness. What is "real"? Real is interaction of the F2F (face-to-face), everyday kind, the kind idealized in the movie set in Oregon, which could have been taken from my youth, *Stand By Me*.

Technology drives much of this. And advertising. And capitalism. Texting is big business for the phone and cellular companies. If we build it, they will come. If earlier generations had had the Internet and cell phones, we might not have spent time outside, walking, playing, chasing girls and boys. We might not have skipped school to join a peace or civil rights rally. Those were our hyperrealities, the times we felt electric and out-of-body. I'll never forget marching on draft resistance day in the late 1960s; I can still smell the patchouli oil worn by my sometime-girlfriend and I can remember the risks we took skipping school. Then again, Vietnam beckoned. As did Canada. We had much to lose if we stayed in school and did our algebra homework for that night.

As an adult, I didn't own a cell phone for a long time because I didn't want to be owned by it. Eventually, I inherited my son's old BlackBerry and realized that texting is a kind of tethering, especially to my kids. My routine is predictable; my wife and kids usually know where I can be found—and available by landline.

I spend much of my time alone, writing and reading. As Hannah Arendt once said, I am never less alone than when I read or write. I project myself into the worlds of others and they enter mine. Generational boundaries blur and I, too, use email and surf the web. Most academics couldn't work without email, which, compared with texting and tweeting, involves slow writing and reading. I use the web as a research tool, even though there are downsides to the thin discourse of Wikipedia and other such sites. I don't particularly want to talk to people by phone when I am on the move, and I am far too clumsy and verbose to text. I quickly exceed the number of characters allowed. I am also frustrated because I hate the back and forth that could be sped up if I just picked up the phone and talked. I know that I am showing my age. Friends and colleagues text and even use social-networking sites and they don't catch the plague!

The problem is partly technological; I am a nature boy and I worry about being ruled by the computer or the telephone, just as we are already ruled by the car. I am probably ADD or ADHD and never knew it. Or perhaps we all are in the sense that no one is really adept at sitting still, especially when the outdoors beckons. I calm myself by running, a better outlet, at least for me, than the cell or computer. Perhaps my utopia is somewhat Luddite, although, as a modernist, I agree with Marx and Henry Ford that mass production is the only real answer to global poverty and need. I don't want to grow my own food, although I have in the past. Mass production has been transcended (negated, preserved, synthesized) in a post-Fordist era of just-in-time, grow-your-own production, where goods are demanded and destined for niche markets. Small is beautiful, but only if global needs for subsistence have been met, something we are far from accomplishing, given the concentration of wealth in capitalist countries.

Technology as Domination, Technology as Play

The interplay between technology and capitalism is an old story for social theory. Marx and Weber addressed these phenomena, in the mid- to late nineteenth century. Both viewed technology as potentially utopian, but only in the sense that it could free us from want. Not until the 1960s were nondominating "playful" technologies and sciences speculated. Marcuse in *Essay on Liberation* (1969), built on his earlier work *Eros and Civilization* (1955), postulated a play impulse that would lead people to be hippies but also to be high scientists and engineers, playing with concepts and techniques much as musicians play music and athletes play the body.

The cellular technologies that define life for kids and adults today could go either way. It is easy to view computers and cell phones as imposing their logics of command and control on a pliant public who are just keeping up with the Joneses. It is also possible to view such technologies playfully. They could be bent to human needs, defined and designed to facilitate the need to communicate, to be heard, to acquire and develop knowledge, to write and touch the soul. Foucault offers one perspective on a sociology of the digital, Marcuse another. Foucault stresses the disciplining and punishing dimensions of having to check your phone and email, while Marcuse stresses the playful, democratic, even erotic dimensions of having the world at your fingertips.

Sociologists since Ogburn (1964) recognize the dual nature of technology: It can be used for and against human purposes. Indeed, the real question is perhaps "whose purposes?" Apple contends that you cannot live without a MacBook, while I tell my kids that a low-end PC would be much cheaper! When I go with them to the computer store, I am always dazzled by the Macs, but not to the extent that I believe they have a soul. They are tools, but tools without which I couldn't work effectively. And work is my play. Similarly, I need to replace my not-inexpensive running shoes every four months or so lest, at my advanced age, I get shin splits, stress fractures, or worse. Here, my play is my work. One can economize, purchasing computers and running shoes from after-market vendors online or, in the case of running shoes, loading up on models out of production. Kids, like most consumers, are suggestible and they believe that the iPhone or Mac or Xbox will change their lives just as adults believe the same about a Lexus or HD television. Part of our technology dependence is a sheer outcome of advertising, created by profit makers to fuel competitive spending. However, these same technologies afford opportunities for self-creative play that trump the issue of their cost or the condescending advertising campaigns that tell us we cannot live without them. Of course we can live without them, but perhaps we don't want to.

An adequate sociology of children's writing must contend with these dual realities. It should not only identify kids on the social-science view screen but examine their use of technology. The Frankfurt School, with Foucault, recognized that technology often embodies a logic of domination—over the self who uses it and over nature, which is its victim. But who is doing the using (even ignoring the drug-era double entendre)? Are the kids using Facebook, texting, and Twitter or are they being used by them? As with all interesting philosophical questions, the answer depends on context, notably on how aware kids or for that matter adults are about the interface between technology—which imposes its own language game, constraints, and imperatives—and human free will. The more aware we are of who is doing the using—and for what purposes—the more we can control and even play with the technologies. We can share our literary wares and build community globally and instantly. What a huge leap forward, as long as we do not fall under the sway of digitality's compulsions. In other words, the more we theorize the digital—think clearly and structurally about it—the less we will be used by it. This was Aristotle's posture, originally. Know thyself. And also find a balance among one's passions. Still true.

The balance may tip in favor of technology as play (rather than as domination) especially if the user has something to say—if he or she has a self, in other words. Too often reporting about life, on blogs and Twitter, has replaced the living of it. You cannot grow a self apart from experimenting, learning, dialoguing, writing. Most of what is out there is depressingly prosaic, and I am not singling out children. Adults' blogs are often banal. Twitter captures this: With a parsimony of keystrokes, you can tell others where you are and what you are doing. This is cyber cinema verité or a lived ethnography. If living becomes all bound up with journaling about it, then there will be no living apart from the journaling. Life provides the blog its fodder, and where blog leaves off, living begins. And living can include more formal literary postures, such as writing for publication.

You have to make your share of mistakes, which definitely include putting your life on the screen for all to see. The un-ironic self-revelation of the social-networking sites is like the tattoo you got in college when you were drunk, on a dare. It stays with you, but also fades. You will look back on your literary adolescence and cringe, but also remember the good times, such as the friends you made and the fun you had while writing.

Does Format Matter?

In many ways, this is the most literary of ages—and of generations. Digitality allows a globality of self-expression, of the written word. Not every self-expression is elevated; pixelated words can be tossed off effortlessly and forgotten just as quickly. But they are different from hallway conversation in that they are more deliberate and they are composed behind the shield of the screen, which mediates interactions in ways that face-to-face conversation does not. You cannot see your interlocutor. And you can pause and consider your response. And it is easier to posture and spin.

When digital technology is approached in this way—as play, as opportunity, and not just as competitive consumerism—it can be a preparation for real writing and real life. By real writing I don't mean just paper printing; I mean writing that is deliberately composed to make a point, to advance an argument, to cover the subject. Carefully composed prose like this has long been the domain of printed books. They tend to generalize and build incremental arguments toward a larger thesis, a structure that may be more suitable to booklength works than the unit format (one blog post a day, for example) of blogging. The difference between this book and a blog or a text is that it generalizes, even as it builds, blog-like, from the particulars of everyday experience—mine and yours. Books are longer than blogs, although many blogs have archives that stretch back many thousands of words. Blogs are often more episodic, more attentive to the everyday, though even some books respond to what is momentary in culture, like those quickly produced biographies of celebrities who are stars one day and forgotten the next.

Stylistically, blogs tend to be first person whereas books are often written in the voice of academic aloofness, a distinction that is probably disappearing as blogging expands in scope. Perhaps the primary way in which blogs and books might differ is that blogs seem offhand while books contain considered prose. Writers may not self-edit when blogging in the way that they (and their publishers) do when writing books. You toss off a blog—about your cat or your favorite recipe—while you agonize over a book. But in the era of word processing, books can be written quickly and there is nothing that prevents blogs from being carefully wrought.

Another obvious difference is that blogs are serial, with entries added by the sponsoring author and by her respondents. Electronic books could have the feature, too, in the future. Socratic dialogues might be a format for this. Another difference is that books are longer than blogs, although many blogs have archives that stretch back many thousands of words. Word count might not be a decisive issue. A possible difference is that blogs are more episodic, more attentive to the everyday, but, even then, trade books might have the same feature. I'm thinking here of quickly produced biographies of celebrities who can, all too quickly, be shown to have feet of clay. Consider the relationship of Paula Broadwell and David Petraeus, at first biographer and subject and, later, more.

The boundary between booking and blogging fades when we recognize that both are writing, both compose the self, and both are intended to be read, critiqued, discussed. Not every blog meets these criteria, but many do. And kids are staying up late not only writing but also reading—blogs, texts, tweets, social-media postings. But am I right about this? Do blogs intend to be read in the same way that more traditional literary forms do? The jury is still out on this. It could be that blogs are meant simply to be written, in our "attention economy" dissolving harmlessly into cyberspace. Who has time to read them? Perhaps that is precisely how an elder like me misses the point! Blogs grow and multiply. There are people out there with time on their hands. Perhaps this is the post-Fordist workforce, both kids and adults who work outside the office cubicle and the traditional forty-hour work week. I sometimes worry that my image of kids' cyber time usage exaggerates, but then again maybe I underestimate, if only because I retire early in the evening, leaving nighttime to the vampires. Many books also go unread or are at best sparsely read. This is especially true of narrow academic works, for example published dissertations that explore highly specialized fields. The blog/book boundary is more flexible and permeable than might meet the eye.

Of course, format does matter. Much of this book is about boundaries—how they are deceptively rigid, not adequately capturing the fluidity of reality, how they shift, how sometimes we need firmer ones. Perhaps the central boundary under examination is between the generations, those who grew up with the Internet and those of us who didn't. Technology intervenes to divide generations. We boomers grew up with TV, while my parents came up with radio. These technologies are imprinting, especially in the ways they divide and connect people. My kids, teenagers both, are tied to others and to the world in ways that I could have scarcely imagined when I was their age. I watched three black-and-white channels of television and we traveled a lot, especially to Europe. But I couldn't "be" anytime/anywhere, nor could I reach out to my friends and other peers as effortlessly as my kids do. Above all, I couldn't leave my traces for all to see by my teenage writing, in effect creating who I "am" or "was." I spent much more time in the library than my kids do, and my approach to books is somewhat different.

Facebook boundaries kids' worlds from their parents. It is adolescent cocktail party chatter—gossip, self-revelation, flirting, "hooking up" (Bogle 2008). Kids are taking back the night, when they post and chat. They are seizing the power of the pixel, even as they are controlled by these electronic media in ways that they don't always understand. This book seeks to explain this to our busy literary children, risking the condescension of adults who are always trying to explain things to kids. At the same time, it seeks to let kids speak for themselves, to show adults that their literary work is a much broader and inclusive activity than we imagine. This helps reinforce my view, formed as a parent, that democracy must be broad enough to include kids, who need their own voice.

In the next chapter, I look more closely at three important features of kids' secret writing, as they connect quickly yet without disclosing their innermost selves. This is necessarily a political reading because adults hold most of the power; kids constitute an important minority group and, like other adult minorities, they often resist.

CHAPTER 3
READING KIDS' WRITING POLITICALLY

The decline of the book, of the considered text generally, is matched by the rise of Internet writing, phone messaging, and video games. College students hunker down over their laptops, composing email and instant messaging. They text and tweet on their phones, which double as computers. Students don't read newspapers, but get their news via the Internet. They are worldly, but they don't know very much, at least by pre-pixelated adult standards.

It is tempting to blame the young for what they don't know, for not being more like us—their activist boomer parents awash in a traditional textuality and political (even theoretical) discourse. But blame is beside the point. I argue that children are the most invisible minority, ignored by "adultist" adults who cannot fathom that kids live secret lives that include both writing and even political imagining. Kids "prefigure" a world in which adults don't accelerate and administer their lives by dealing with each other through plural and democratic discourses that rely heavily on rapid electronic media. Perhaps this is my guiding image: Kids, almost all of whom deplore school as a prison, sneak in their cell phones and text each other during the school day in order not only to let off steam but to achieve a democratic community in which no one assigns homework. This is to view kids as belonging to a pre-labor force, constituted by adults who resent and envy kids' primal freedom, and also to view them as utopian actors, although defying detection by traditional social theory and social science.

Texting, social networking, and blogging constitute *prison talk*, tapped out in code available only to initiates and oriented to building countercommunities. This might be viewed as political if it is in the service not only of busting out of but transforming what kids experience as prisons (schools, polities, economies, families). Kids' prison talk is pre-political, potentially political. We need e-sociologies that learn from the traces of these electronic emissions about how kids are experiencing their own lives.

Here, I consider writing and reading via the Internet and mobile devices as a dialectical phenomenon, containing both positive and negative aspects, aspects that interact and conflict with each other, perhaps combining into a higher synthesis. On the negative side, libraries are either dying or becoming coffee and computer lounges. Pulp academic journals are going online. E-books abound. My students don't read much of their old-fashioned assigned pulp books or articles, but they download from the Internet, chat, text. They are at once immersed in words and impervious to certain types of words—contained in dusty tomes hidden away in libraries. This provokes images of Bradbury's *Fahrenheit 451*, a dystopia in which books are banned and most survive through memory and are circulated via storytelling.

On the positive side, the book as a writing format is on life support, but long live the furious keystroking taking place among generations X, Y, and Z. And I hesitate to defend the pulp past as a golden age of wordsmiths and careful readers. I hated grammar in school; the great books were rarely great and they were exclusionary. I rarely took an English class that sparked me. And there are certain things I love about the Internet—its indie, edgy quality, its resistance to the commodity form, its inherent democracy. I am a cyberdemocrat (Kann 2005), but I sense that life lived on the screen frequently does not serve the project of democracy well because it thwarts community and attenuates reason.

Discourse in the Age of Fast Capitalism

The dialectic of discourse in the age of the Internet and mobile phone takes place within a fast capitalism in which culture and communication industries exist alongside of, and reinforce, traditional industries that provide goods and services. We still inhabit capitalism and have not transcended beyond it toward a putative postmodernity. Postmodern theory refreshes critical theory where it draws attention to the importance of culture and discourse in reinforcing domination, the argument first made by Horkheimer and Adorno when they proposed the concept of the "culture industry." But they were not departing from Marx here: Cultural domination only extends the alienation of labor by diverting people's attention from their miserable lives and by opening huge new markets for entertainments, goods, and services unimaginable in the penurious era of Adam Smith's original eighteenth-century paean to market capitalism (2003 [1776]).

Fast capitalism is a late-twentieth- and early-twenty-first-century version of capitalism in which production, consumption, and life generally are accelerated. There is little down time. Too much time would give people room to move and to think. Electronic prostheses, from radio and television to the Internet and mobile devices, build us into a disciplinary grid. This effaces the boundary between public and private, in effect colonizing not only the household but the body and psyche. Compulsively checking one's email or mobile for messages is only one example of this colonization.

Undoing fast capitalism certainly involves slowing things down, rebuilding the private sphere, family, body, and psyche. This will be a discursive solution in that people will take back the text—now many texts, from pulp to pixels and now including images, music, and film. Early Marx portrayed utopia as a society of praxis, activities that define people and allow them to express themselves. I contend that writing and reading are two examples of self-creative practice, early Marx's praxis. We live in contradictory times: people yearn to write busily, composing emails, tweets, and text messages, and yet these nontraditional forms of textuality are also "language games" that thwart the self in significant ways. In this chapter, I examine these contradictions, preferring to view them not as frozen but as dialectical, possessing the potential for their self-overcoming.

Uncoerced Instantaneity

Young people write in order to make connections, an opportunity opened up by a fast society in which information and communication technologies, originally designed for the space race to the moon, are accessible to many. They want to connect, and now, not via slow letters or, even worse, books and articles that might take a year or more to see the light of day. This desire to write in order to forge community is a genuinely utopian impulse, even if it does not fit the framework of traditional academic writing. My college students would much rather chat, text, and tweet than write boring term papers, which require self-directed research and not memorization, to which they are accustomed. We need to recognize the motivation for this and not simply dismiss it as yet another example of our (baby boomers') generational superiority to the post-sixties generations.

If, as I contend, people are bursting with literary intentionality precisely because they feel so disempowered politically—and because, as ever, they are innately poetic and creative—why do they prefer chatting, email, and texting over old-fashioned pulp literary craftsmanship? There appears to be a cultural attention deficit disorder in play, although I maintain that ADD here is symptom and not cause. We, especially the young, favor quick and unstudied discourse for two primary reasons: Our lives are already fast, clogged with work, family, school, commuting; and we are awash in a media culture (Kellner 1995) in which the sound bite and emoticon are more available to them than are the process and product of studied literary craftsmanship. This is not necessarily a value judgment but simply an observation about the interplay of social, cultural, and technological circumstances in fast capitalism. My son does most of his writing on a laptop, which requires and reproduces certain technical skills including the ability to do graphics. But the laptop also requires him to type. Although he composes old-fashioned prose—perhaps reflecting the fact that he reads a lot—he is enticed by the faster discursive formats available to him, but that were unavailable to me as a child and obviously unavailable to citizens of the fifteenth century.

In effect, then, young writers are both used by, and they use, the rhetorical vehicles of what I am calling "instantaneity." They compose quickly, and they expect a quick response. However, it is clear by now that many younger writers, readers, and callers prefer to stand behind the written word, albeit pixelated, and voice mail rather than to talk voice-to-voice, let alone face-to-face in the public sphere, such as it is. Perhaps Starbucks is the new public sphere! But I suspect that the main literary work taking place in Starbucks is via composition on cells and laptops, enabled by the proliferation of Wi-Fi in public places.

Instantaneity is desirable for younger users and for some older ones. It makes possible what the young call "hook-ups," which in some contexts means casual sex. But it also refers to a range of behaviors from getting together with friends to making out to just making contact electronically. One meets, greets, and passes on, without strong anchors in a relationship. Instantaneity also satisfies the desire for immediate gratification, which is not a character flaw but an outcome of a fast economy in which consumers are encouraged to be restless and insatiable. Quick connections make for instant gratification and fast profits. Since World War II, capitalism has shifted from an ethic of savings to one of spending, often beyond one's personal means using credit. Instantaneity is characteristic both of electronic hook-ups and of an Internet-era capitalism, in which consumers are satisfied "just in time." Here, the economy and cultural sensibility blur, just as Weber noticed that they did when the Reformation made way for early capitalism.

The desire for quick hook-ups would seem to argue for the use of mobile phones not only for texting and voice mail but for actual voice-to-voice connections, in real time. But younger people, although they do actually talk together, often avoid these connections in favor of asynchronous, if nearly instantaneous, connections such as texting. This leads to the next rationale for fast communications, which is people's penchant to manage the impressions they give off to others. There is fundamental concealment going on, although, as I will argue, this is less born of the motive to deceive than of a rootlessness characterizing the postmodern sensibility. It is not so much that people choose to manage impressions than that they feel invaded and overwhelmed by actual interaction of the kind exemplified by traditional voice-to-voice phone conversations.

All of this indicates a contradictory quality of fast communication today. People want it quickly, but not so quickly that they have to improvise social relations with absolute immediacy. The boy wants to ask the girl for a date. Texting her takes the edge of his shyness, which would be tested if he actually called her. Youngsters desire improvised instantaneity, making connections in a risk-free, deliberate way. "IluvU" wagers far less emotionally than picking up the phone and saying it out loud, risking rejection or a wandering conversation ruled by serendipity.

Ethnomethodology, beginning with Harold Garfinkel (1967), has ingeniously identified the ways in which people make sense together through conversation, in effect building social structure out of chatting. They don't have a manual of interactional rules, nor do they possess perfect understanding of what is being said to them. They muddle through, interpolating meaning whereas, if the conversation had been transcribed, the exchanges would seem clumsy. Garfinkel revises Parsons's 1951 theory, which traced enduring social structure to certain meta-level institutional processes that bear down on selves, in effect coercing them. Among these institutions he named family, nation, polity, religion, and the like. Yet there is coercion implied in conversational exchanges: You are forced to listen, infer, improvise, and respond. Your partner may yell at you or hang up—or propose marriage!

Electronically mediated communication removes an element of coercion from nearly instantaneous exchanges. You can ignore the text message or tweet. You can chat with someone else. You can withdraw or advance. This can be rather self-absorbed, trading the lack of coercion involved in turn-taking for a lack of empathy, which Habermas (1984, 1987b) regards as the foundation of communicative democracy. For him, as he translated first-generation critical theory into second-generation communication theory, democracy is found in the embryonic conversation: people prefigure a good society by agreeing to take turns, to refrain from browbeating and propagandizing, and to be governed by the power of the strongest argument. This is an old-fashioned (but quite appealing) modernism compared with the uncoerced instantaneity of electronic chatter, which trades obligatory turn-taking and even empathy for the absence of interpersonal care.

The paradigmatic act here is the glance at the phone number of the mobile caller. One can allow the call to roll over into voice mail after a few rings or turn off the phone altogether, triggering an immediate rollover. Or one can pick up the call. Although such selectivity would appear to be value free, there is a hidden normative agenda in caller ID. Knowing the identity of the caller invites nonresponse, which could be seen to be rude (and prefigurative of an uncaring society) or seen as usefully removing an element of coercion, such as having to respond. Instantaneity is especially coercive as people experience the deluge of telephone solicitations, especially at home in the evening, during "private" time.

Responding is essential in a democratic society. It creates norms of reciprocity, mutuality, and accountability. I suggest to evening telemarketers that their interventions are unwanted. Once, I threatened legal action when we were called several times over two days. I actually returned the call and sought the owner of the business. I finally reached her, and she responded that I had no legal standing. I then called a policewoman who was being pestered similarly while at home, and she initiated an investigation into my nuisance caller. This is more effective than simply hanging up on the unwanted caller or allowing the call to roll over into voice mail, which I would have to listen to anyway. My responses built community with the policewoman, with whom I formed a micro-level social movement!

I suspect that young people seek uncoerced instantaneity not because they worry about being bombarded by telemarketers but because they are unsure of themselves and not confident in social interaction. As well, they may not understand a Habermas-like perspective on how democracy springs from what he calls ideal speech. In plain language, they take rudeness for granted, and they perpetuate this rudeness unwittingly as they engage in interactional selectivity through caller ID and Facebook ignoring. My daughter knows a girl who felt snubbed by a boy in the same grade because he no longer picked up her calls. My wife brought this to his attention gently and he shrugged, not recognizing that his inattention was hurtful. He brags that he has the phone numbers of fifty-some girls, a panoply of potential interest. However, dealing with all of these potential hook-ups would be a full-time job for him; some selectivity is in order. As well, this unfortunate young man told me recently that people would soon think he was gay if he didn't have a girlfriend. My guess is that he stopped dealing with the aforementioned girl with whom he was having a platonic relationship because he decided to invest more time in an actual hook-up, thus demonstrating an appropriate sexual orientation to his male friends.

It would appear that communicative coercion (having to respond, formulating a coherent response, being quick on your feet, and so forth) could be reduced further if we eliminated interactional instantaneity. Imagine sending a slow letter the old-fashioned way. Or writing and publishing a book or article. Or tossing a message in a bottle out to sea. One would get, at best, a slow response or no response at all. But young people yearn both to write and to enjoy community, which is why they punch in texts and compose emails and chat. They type furiously because they are impatient with their isolation. But they do not want boundaries to be completely demolished; they want to pick and choose their electronic relationships and they do not want to feel compelled to react nearly instantly to messages in real time. They resist compulsory communication because they are already so compelled in school and work. Many young people want communication and hence community, but only on their own terms.

As well, they don't have the time or patience to compose the handwritten letter, let alone the booklength manuscript. They don't have enough to say. Writing is nearly embodied as one picks up one's mobile and looks for messages or opens one's laptop in order to chat and surf. These technological mediations constitute an almost embodied presence, a sense that one's interlocutor is in the same room or within an arm's length. Mobiles and keyboards are ergonomically designed in order to facilitate this human/machine interface. And it matters greatly to many kids what kind of mobile they use. Not only do they want the latest electronic vehicles but they also want a certain aesthetic. This is because the mobile phone, like the laptop, stands in for the missing humans with whom one is communicating. We can now videochat, but many people don't want to be seen as they communicate. People seek concealment and selectivity in their electronically mediated interactions. Video might seem too much like face-to-face or voice-to-voice communications, threatening their fragile sense of self.

Indeed, the world of information and communication technologies (ICT) is messy and blurry. Kids who text also spend hours on their mobiles just talking with their friends and significant others. Perhaps there is a level of intimacy that must be achieved for them to bare their souls voice-to-voice, in real time. My daughter talked to her former boyfriend for hours on her mobile, even listening to him play and comment on his video games (raising gender inequity issues for me and my wife). But she is rather parsimonious about voice-to-voice communications with other friends. She much prefers to text, explaining that it is easier to escape detection at school when they text. And there is much debate among school administrations and parents about whether kids should even be allowed to bring their phones to school. After the 1999 tragedy at Columbine High School, in Colorado, many parents and their children view mobiles as a necessary lifeline to the outer world.

Impression Management

One of the principles of postmodernism is that the self or "subject" is not stable but open to multiple, often contradictory positionings. This positioning is often thrust upon the self and not necessarily chosen by him or her. Young people accommodate this by, in effect, acting as if their selves are not singular and stable but shifting and various. They thus become adept at what the sociologist Erving Goffman (1959) called "impression management": constructing the presentation of the self in order to engender certain reactions. This is false in the sense that the real self is kept concealed, backstage to use Goffman's dramaturgic metaphor. However, die-hard postmodernists would contend that there is no "real" self, no backstage version who dons a mask simply to perform the play. Instead, the self is always contextual—con-textual, if you will—and thus it has no inner reality separable from the ways in which it performs and presents itself discursively. In short, the self is always highly mediated.

One of the sources of this mediation, which occurs in a managed and deliberate way, is the electronic media through which we connect ourselves to others. Texting and tweeting are vehicles of impression management. There are positive ways in which the plasticity of the electronic self can be useful, especially to young people trying on social relationships. Goffman for his part implied that we always manage impressions, having moved beyond a Hobbesian state of nature. Mead and Freud would probably agree that there is no stable or singular inner core that we could reveal nakedly to the outside world. We always cloak that inner nature in the protocols of everyday interaction. The early social psychologist Cooley described maturity as the ability to view oneself as others do—a looking-glass self.

We understand well the awkwardness of adolescence, when we worry about our complexion, our clothing, our car. The worries ease as we come to care less, perhaps because we can more reliably predict how others will view us. And yet there is always a distance or disconnect between the inner and outer, as psychologists have long recognized. The management of this distance creates impressions—choosing the right jacket to wear on a job interview, bringing the right gift to the parents of your date, moderating your views in a committee meeting so as not to appear over the top.

The boundaries around the self began to dissolve, perhaps, with the emergence of a media culture in the 1950s and 1960s. The Frankfurt School talked about the decline of subjectivity, of selfhood, and the eclipse of reason as concomitants of a late capitalism. My discussions of fast capitalism take a similar tack, although I trace this decline to what I call the eclipse of the book, of textuality. Now books ooze out of their covers and into the world, commanding people to enact their encoded entreaties without having to be didactic. In an electronic- and now cybercapitalism, texts flood out of their pulpy jackets into the vast sea of pixels and images inundating us from morning to night.

We must disentangle the ordinary pains of growing an adult ego from the extra or "surplus" pains of growing up today, when there is so much transparency in human relations and electronic contact. Our media culture involves total surveillance, based on self-surveillance, in that we imagine that we are always being observed, especially in the prison-like environment of the school and by mistrustful parents. Kids are overburdened by peeping parents and teachers, and they are put to work, churning out homework and other performative (evaluated) activities in their after-school activities, from sports to music. They are not only being trained for a productivist adulthood; they are also objects of adult envy, who wish that they, too, could enjoy the carefree days of childhood, given over to "purposively purposeless" play.

In this context of too much to do, too little time to do it, boundaries that used to separate social institutions and ensured a barrier between public and private life have come tumbling down. This experience of boundarylessness, which could be another way to talk about globality, is quintessentially postmodern, even as we continue to inhabit a capitalist economy. Marx predicted both globalization of capital and the postmodern experience of a virtualized reality that shifts and permutes in weird ways. He and Engels in *The Communist Manifesto* (1998 [1848]) characterized what I am calling the postmodern experience of a deboundaried existence as "all that is solid melt[ing] into air." People lack a stable foundation, including a psychic foundation in a stable identity. All of life becomes a stage because we feel that we are simply playing roles and concealing whatever authentic inner experience remains.

Modernists and postmodernists differ over whether it is still useful to talk of a "self," an "identity," or a "subject." I take the modernist view that we must work hard to protect the self, even as we acknowledge that the self is increasingly imperiled by the demands of impression management. In other words, the self is a historical question: how much of it there is, how much autonomy and agency it possesses, and its potential to link with others (not the hook-up described above!) in changing the world. Hard-core postmodernists such as Foucault (1977), Baudrillard (1983), and Lyotard (1984) basically disqualify the self or subject a priori; the Frankfurt School treated the subject as a hypothesis, to be resolved empirically with reference to the earlier questions about agency and efficacy.

Kids find communicative technologies tailor-made for managing impressions, to impress not only adults but their peers. Adults, too, manage impressions. Consider the investigative TV show that depicts adults posing as kids online, striking up conversations with adult men, and then agreeing to meet these men in person. The men make arrangements to meet the alleged children and are instead greeted by the television crew or police. It is surprisingly easy for the adults to pretend to be early teens naïve enough to pursue sexual relations with these adult men. Likewise, the older men pursuing a sexual relationship pretended to be younger and more virile than they really were. The show is a hit. Viewers cannot believe the perversion of seemingly normal adults (teachers, lawyers, men of the cloth) who would seek sex with children and then be dumb enough to show up at the house being used for the sting. Once the men are apprehended, they are led outside, where the police force them into prone positions as they search them for weapons. This is followed by degrading debriefings by officers who entice the men to talk about themselves and what they were intending to do with the supposedly teenage girls or boys. This television show is a paradigm of how strategies of impression management are various and how they sometimes work at cross-purposes.

Electronic communication technologies, to the extent that they permit asynchronous (lagged) interaction, are perfect for the management of impressions. I am thirty-five and skinny, when I may actually be fifty and portly. I am old enough to drive and drink, when in fact I am not. I am old enough to have sex, when in fact I'm not. I'm single, when in fact I'm married. I'm married, but in fact I'm cohabiting or getting a divorce. My dog ate my homework, when in fact I slept in and missed class because I was hungover. One could make some of these same misrepresentations face-to-face or voice-to-voice. However, it is easier to dissemble using lagged electronic communication such as texts and email. One can use paper media to manage impressions and tell downright lies. One could present a work as one's own, when in fact it is plagiarized. One can adopt a nom de plume. One can present journalism as one's own, when in fact one used unattributed sources. However, impression management works better where it is ephemeral and in the moment. One frames oneself quickly, flexibly, in response to shifting external stimuli. You present yourself one way to one person (for example, online) but differently to someone else. You must keep track of these (mis)representations: Everyone knows someone, perhaps oneself, who sent an email to the wrong person, provoking hilarity, embarrassment, or both.

Impression management has a cynical side as well. "Personality salesmanship" is, in the American context, born of a busy entrepreneurialism that requires people to sell things, including themselves. Marketing is its own college major that trains people to manage impressions in order to make the sale. Goffman is not celebrating a business culture but noticing the disjunction, and sometimes outright contradiction, between people's inner and outer natures. Mead talked of the distinction between the I (the interior, "real" person) and the me (the public persona). Freud talked about the id and ego (mediated by superego or conscience). As a neo-Frankfurter, I regard the distance or difference between inner and outer natures as political, thus subject to historical variation. The freer the society, the more people can be themselves, saying what they want, doing what they want, wearing what they want. The less free the society, the more people must don masks and play roles.

A good Derridean already recognizes that the positivist goal of perfect, static representation is a myth. The knower is always already implicated in the world he or she is trying to describe; hence, there is insufficient distance between subject and object for clear representation to take place. Language, including that of science, is, to borrow from Nietzsche, a prison within which meaning is locked up. This does not mean that we should not try to clarify things, but only that clarification is an endless and communicative process that turns science into a fiction. There is nothing wrong with science fiction as long as science confesses its own rhetorical nature, its undecidability, to use Derrida's term. And so we should not take an antiscience posture but rather view science as one literary attempt among others to talk about a world in endlessly ambiguous language within which we are always already ensconced.

One needs to distinguish here between language, which is shifting and elusive, and the self, which may or may not be shifting and elusive. Diehard postmodernists contend that the self is a fiction because science and other narratives are fictive. This is not my posture. This is not to say that the self is singular and fixed, but only that it is relatively more stable than the postmodern concept of the "subject position" may imply. The self has the *potential* of stable meanings and internal continuity; indeed, that might be one definition of good mental health. Sometimes, though, the center does not hold and people begin to unravel. Whether this makes them crazy or insane depends on the psychiatric discourse one favors. I tend to agree with the antipsychiatry movement (Laing 1967; Szasz 1973) that a great deal of psychiatry is simply punitive labeling that actually reproduces its object both by stigmatization and through the self-fulfilling prophecy.

Although Adorno (1973b) condemned Kierkegaardian existentialism for its "jargon of authenticity," Sartre (1956), later to become a Marxist, distinguishes in *Being and Nothingness* between acting in good and bad faith. Acting in bad faith is where you pretend that your actions were actually not willful, perhaps the product of social laws, peer pressure, or mental illness. Acting in good faith means that you take responsibility for your actions, which you freely chose. Indeed, you acknowledge that you are an agent, a producer and self-producer of both yourself and the world. This agential concept of the self is very important to the existential-Marxist tradition that can be traced to the early writings of Marx, where Marx said that the self is a creature of praxis, self-creative action. Whether or not one considers the self "authentic," Sartre clearly wants to encourage us to view ourselves as processual, unfolding, indeed undergoing a process of construction and—here is where I come in—composition. We write who we are, and thus become fully human. This Sartrean/early-Marxian vision of self-creative practice humanizes work and argues that we become human through work, which, borrowing from the Hegel of *Phenomenology of Spirit*, involves our self-externalization in nature.

Depthless Discourse

If there is a literary drive or impulse, in the early-Marxian sense, it plays out in fast society quite differently than it did when I was a child growing up in the late

1950s and early 1960s, not to mention than it did during the Middle Ages. We had no answering machine. If we missed a call, we did not know about it. If I wanted to write my girlfriend, I had to compose a note and give it to her at school the next day or perhaps write a letter that I sent through the mail. I could always "send" the message via gossip networks, asking my friend to tell her friend and so forth. Now, writing is composed hastily, both because commercial publishers pander to the public (and thus reproduce it) and because fast technologies such as tweeting and texting promote thin, terse discourse, not the thick, labored kind. Chat is composed in computer quick, laden with emoticons, acronyms, and shorthand that compress meaning into a few keystrokes confined to a single line of text, or perhaps several, but with nothing as weighty as a paragraph. Texting is similarly constrained by the available technology: One must hunt and peck a quick message, constrained by a small screen and the cost of messaging.

These fast discursive technologies are themselves socially situated. I am not proposing a technological determinism but noticing that the technological framings of discourse in fast capitalism are well suited to the postmodern sensibility that prizes unconstrained instantaneity and impression management. These social and technical contexts frame a depthless discourse in which nothing much is said, but quickly. "RUready?" "Lol," "luvU" replace the thick phrasings of an earlier era of pulp, distance, authenticity, thick discourse—a world of landlines, slow mail, and even fountain pens.

Serendipity and even sheer accident play roles here. The miniature computer chip came from the US/USSR space race as the engineers realized that they needed on-board computing to enable the spacecrafts to process data quickly. At that time, computing was restricted to cumbersome mainframes, which could occupy a whole floor of a building. From the computer chip came microcomputing—desktops, laptops, and now smartphones. The Internet made use of these microtechnologies even as it had a somewhat different genealogy: It was put into place in order to allow governments to communicate in case of nuclear war. Fast-forward to the present. Kids use their mobiles and computers in order to "do" discourse. They are not traveling to the moon, but it might seem that way to them as they navigate the strange, stressful worlds set up for them by parents and teachers. Texting, tweeting, and chatting are their ways to stay connected, but they are also ways for them to create a kids' world off limits to adults.

This is a virtual, simulated world, which also includes video games. These games, like the electronic communication discussed here, simulate reality but of a playful, exciting kind where there is risk and gain to be had. School is universally condemned as boring, with standardized testing, rote homework, and authoritarian classrooms. Exciting is gaming, texting, chatting—all venues for the simulation of other worlds and other selves. I contend that underlying all of this is a creative impulse to imagine the world differently, indeed to compose that world, which is also an act of self-composition. This imagining is a utopian impulse, and thus potentially a political one.

Although it is easy for adults to dismiss gaming and texting as depthless—and the emoticon-riddled world of computer quick discourse is nothing if not superficial—it is also a revolt against time robbery. In Chapter 8, I consider ways in which people, especially kids, create alternative realities unconstrained by synchronous time in order to transcend, and to protest, their regimentation in the sweatshops of childhood and adolescence.

Kids who pursue community through uncoerced and instantaneous electronic communication are literary and political rebels. They use words to create their own intimate worlds. They rebel against their domination by a forbidding adult world. This can be rude, where monologues prevail. And it can render the self plastic, spread too thin and always posturing. Sixties lefties like me are old-fashioned enough to decry the depthless discourse of the chat rooms, instant messaging, texting, and even email. We prefer wordy screeds, composed contemplatively and in their own good time. It would seem that time rebels such as our gaming and texting kids would want to slow down their lives, not give in to instantaneity. But they are starved for community and connection, and they have the technological means to create it, or to attempt to do so. I am ambivalent about the Internet in this sense: it is nearly frictionless (for those who can afford it); it is democratic, amateur, and possibly edgy; and it makes way for cyberdemocracy and hence a new public sphere. And yet so much of what is out there, both on the web and in these media of electronic chatter, is insubstantial, amounting essentially to self-absorption or simply commercialism.

Engage in a thought exercise. The young Michigan students who in the early 1960s started the Students for a Democratic Society (SDS) that eventually stopped the war in Vietnam, led the War on Poverty, and empowered a whole generation who rallied around the cry for participatory democracy—the centerpiece of what they called a New Left—met at a summer camp on Lake Huron, Michigan, in the summer of 1962. Their literate leader and scribe Tom Hayden brought to the meeting a draft of a manifesto for this New Left. They spent several days debating and revising the document. It was issued as *The Port Huron Statement* and was mimeographed, stapled, and circulated for a quarter

(Hayden 2005 [1962]). It made the rounds for the rest of the decade, dog-eared, annotated, with pages missing. It changed a generation and the world as it convinced young Americans and youngsters abroad that Camus and early Marx could meet on the terrain of existential/political commitment.

What would have happened if Hayden and his cohorts had the Internet then? On the one hand, they would have achieved instant globality, spreading the message overnight, or sooner. On the other hand, the document might have been swallowed up as Internet flotsam and jetsam. It would have been drowned out by spam, advertisements, iTunes, email, messaging, and the other electronic claims on people's attention such as television, videos, and games. *The Port Huron Statement* could not have been tweeted; it had too many characters.

Sociologies of the Virtual

Writing in the age of the Internet has a dialectical quality. It reflects a rushed society in which intellectual life has been reduced to text messages, composed and read hastily. But the busy literary work going on, especially among younger generations, may reflect a yearning for a healthy public sphere and a desire to unleash a pent-up textuality—the urge and urgency to compose the world, thus composing the self.

Sociology needs to address this dialectic, or perhaps better said this ambivalence. The fate of an Internetworked fast capitalism is uncertain. The Internet is a contested terrain: the corporations want to restrict access to paying customers, while open-source progressives, much like the *Port Huron* generation, want to throw it open to all comers, all writers and readers. The Internet is obviously a powerful vehicle of shopping and selling, but it is also an outlet for independent thought and original creation. It can enervate, but it can also educate. Sociologies must wrestle with this interesting conjuncture of the social and the technological, focusing especially on the ways in which electronic connections are sinews of new types of community and vehicles of new forms of self-expression. But what is dialectical about the Internet is that these types of communities and self-expression feed back into the political as they form and forge what Gramsci called "counter-hegemony"—new forms of everyday life that, by their very existence, constitute a kind of refusal.

We are light years away, or so it seems, from getting young people to understand their literary praxis in political terms. They game and text, while we fret and lament. The clock cannot be turned back to Port Huron 1962 or Paris 1968. However, we can use *Port Huron* as an example of an amateur document that, in its accessibility and relevance to everyday life, took on a political life of its own and facilitated civil rights, the antiwar movement, and the women's movement. Such amateur documents need to be written and circulated, this time via the Internet. It is difficult to imagine what will politicize the young auteurs and gamers of today. There is no military draft in the United States. Most young people did not protest the war in Iraq and Bush's shenanigans. They bop along to their iPods, with mobiles in hand. In this context, an engaged sociology of the virtual needs to rethink its old assumptions about political change and social movements, perhaps recognizing that radical inspiration in the future will come less from manifestos and more from images of an alternative everyday life, a decidedly utopian envisioning of new worlds. The kids today do not read in our traditional ways, combing through Tom Paine or Tom Hayden for ideas about how to change the world. They read and write, but not in traditional polemical forms. It needs to be explained to them that virtual media, of the kind with which they are intimately familiar, can not only help them escape but conjure whole new "virtual" worlds that become reality as they are lived. This isn't just the stuff of utopian fiction, but a quite materialist perspective on the possibilities of texts in our virtual stage of capitalism.

Kids love the video game The Sims. Using the game, they manipulate life-like characters as they live their "real" lives. At first glance, it seems boring, at least by comparison with more dazzling video games. But there are endless possibilities for The Sims characters and for the "authors" who invent whole new lives and worlds using them. In the case of The Sims, the virtual and real worlds seem quite similar, especially given the life-like characters leading quotidian lives. I think they love the game because they want a whole new world for themselves, and they want lives of which they are in charge. Virtual reality is not only an escape; it is utopian, and it reflects people's desire to be in charge of things. It is utopian in the sense that it simulates a better world. The line between escapism and utopianism can be quite tenuous. I am not denying that many kids seek merely escape. It is interesting that older people, too, like The Sims, just as they have migrated onto Facebook, which began as a Harvard dating site. Generational boundaries blur, perhaps suggesting that adults, too, seek escape and utopia in fantasy worlds.

However, with guidance, they would not only problematize their own lives but recognize, through their gaming, chatting, and texting, that they could repair and improve those lives—their own. The virtual segues into a "real" reality where people, especially young people, use it as a utopian spur to social change, perhaps beginning with personal change. Again, I am reading the virtual lives of texters and gamers as both protest and possibility, if not yet as a full-blown mode of social theory and political critique.

A good example of a utopian virtuality is the hoopla surrounding the series of Matrix movies. Intellectuals often read into these movies a sense of postmodern plasticity and flexibility, which can have political overtones (Irwin 2002). In this sense, the Internet connotes the hopeful plasticity of our lives, which otherwise seem humdrum and locked in—precisely the impression produced by what Marx in the nineteenth century called ideology. Even if virtual plasticity is not the real thing—better job, love, health—it provides hope where otherwise there might be none.

Why do we often miss the busy textuality underlying youth cultures? This insensitivity to the discursive practices of the young might be said to be born of *adultism*, a systematic oppression of children and adolescents. I and my coauthor have considered the causes and consequences of adultism in our recent book *Fast Families, Virtual Children* (Agger and Shelton 2007). We view children as the forgotten minority, whose lives are being accelerated and disciplined in ways that extend Foucault's disciplinary society. We examine family (actually, families—plural) and education as sites of potential utopian practices, a framework I am using here as I try to read what the young are saying and writing, even uncovering a utopian desire to live in a world apart from the hierarchical and bureaucratic world of parents and teachers.

While young people's writing can be read politically, most kids are apolitical or, perhaps better, pre-political. The 2010s may change some of that as a new coalition of young people, minorities, and women are coalescing to promote a new social agenda that includes the protection of reproductive rights, same-sex marriage, and the legalization of marijuana. This is in contrast to their boomer parents, who were activist during the 1960s and who also inhabited and still inhabit a bookish culture. Can text messages be read as manifestos? These are the issues we'll examine in Chapter 4.

CHAPTER 4
THE PULPLESS GENERATION(S)

The Decline of Activism

Activist parents who were young people in the 1960s often find themselves with children who are largely apolitical. This is an astonishing result for them. Who would have predicted this sort of generational rupture? They are less activist but now have electronic prostheses that make cyberdemocracy possible. Whether this is "real" democracy depends on whether "connection" amounts to "community" and hence social change. To the extent that young people are apolitical, it is partly *our* fault, the elders, because we have offered few useful models of activism in the twenty-first century. It is partly nobody's fault but simply an outcome of a bookless, depthless culture.

By 1968, millions of Americans had taken to the streets in order to protest the war in Vietnam (Wells 1994). When is the last time you witnessed, or heard reporting about, or participated in a massive demonstration protesting Bush's adventurism in Iraq and now Obama's in Afghanistan, which he promises to end by 2014? We confronted the draft then, especially those among us who were draft-age boys, with our lives on the line. But many millions of young women and older men and women also protested the war. Today, all is quiet, or nearly so, on the western front. And yet Bush's job approval rating as president was about twenty-some percent, Watergate era–like numbers. People hated him, but they did nothing about it, at least until they joined and supported

Obama's campaigns for the presidency. Dialectically, though, dislike of Bush has morphed into demonization of Obama, as the religious Right, even in the 2012 presidential campaign, portrayed Obama as a Kenyan socialist outside of the Judeo-Christian mainstream.

Many argue that it was the televised war, with body counts and graphic color images of gore, that turned the domestic tide against Nixon, McNamara, and Westmoreland. Television certainly radicalized American citizens, bringing the war home for them (see Gitlin 2003b; Varon 2004). But television then and now are quite different. Television was, in effect, almost a pulp medium then: It was slow, analytical, pre-graphic. And it was situated within a pulp culture of books and journalism by comparison with which it was a drop in the psychic bucket. Today, television is an almost old-fashioned vehicle awash in a media culture that dominates consciousness. Increasingly, the only television that post-boomers watch is reality television, which blurs the boundary between the representationally real and the reality of lived experience, albeit a lived "reality" heavily mediated by electronic interfaces such as the Internet and cell phones.

Today, people, especially younger ones, are overwhelmed by electronic media that are instantaneizing and oversimplifying, even as they purport to reflect (thus secretly constituting) reality. A print culture, in which writing and reading are mediated by careful thought (reflection as in reasoning and not reflection as in positivist mirror images), is slipping away. The eclipse of the book (and journalism) is what is causing young people—and not only young people—to stay home. Only books and writings provide *context* (con-text) that helps people make sense of their lives in terms of large-scale structural forces.

Jacoby in *Last Intellectuals* (1987) and I in *Decline of Discourse* (1990) analyzed the decline of public books in terms of various structural factors such as the demise of urban bookstores, chain bookselling and vertically integrated trade publishing, and the rise of academia (that swallowed the New Left), with its restricted linguistic codes. And there is also the media culture that abbreviates our attention spans. Before chain bookstores, academia, and the Internet, "big" public books could change our minds and energize our activism. Social critics learned and taught the world using pulp vehicles. Again, I want to avoid a technological determinism that is in effect Luddite. It is not the pulp page per se that elevates thought and facilitates critique; one can use the Internet to circulate wordy and passionate texts. It is a bookish culture, in which people wrote and read slowly and considered arguments carefully that allowed people to be citizens—interested in, and engaged by, the public world around them.

The issue, then, is not simply that kids don't read or write in the traditional ways. It is that they lack the public world in which ideas circulate and causes are formed. There are no manifestos because no one would read them, even if they were to be widely circulated. Books—pulp, generally—could have little purchase today, even if their authors wanted them to. Their avoidance of the book is an avoidance of the public world and politics at large, a world that books helped create. They also avoid ideas, abstractions, arguments. Instantaneizing, depthless writing and reading certainly prepare the ground for the decline of the public world. They contribute to depoliticization. But these are symptoms, although symptoms that double back and reinforce the decline of the book and hence of the public world. Cell phones in themselves don't depoliticize kids. Kids use them because they are already depoliticized. And a crucial factor in their depoliticization is the rise of a media culture in which books are diminished into trade potboilers that have short sentences and that go out of print nearly as soon as they are published. There is definitely a dialectic of ideas, discourse, and technology at play here. Single-factor reductionism is to be avoided.

My argument is that the decline of pulp literary vehicles reduces young people's passion and denies them analytical frameworks for understanding their lives and the world. Ironically, this pulpless generation was born to sixties parents who learned the world from pulp sources and from television. I am not assigning generational blame but attempting to understand the decline of public and political discourse among young people as a technological phenomenon that closes off traditional political vehicles but opens up new ones.

Todd Gitlin (2006b) recently argued that "the kids are alright," by which he means that older adults, especially those who were activists during the 1960s, shouldn't fault the younger generations for their apparent lack of political activism and efficacy. Instead, he argues that some of them are "netrooters," single-issue activists who use the Internet to raise consciousness and organize. He notices that few of them have a sweeping theory that helps them frame their narrower causes. Although I disagree with his contention that the kids today are, in effect, reproducing sixties activism but in local ways and shorn of doctrine and dogma, I share his view that cyberdemocracy is certainly a technological possibility that can in important ways use the means of globalization against its globalizing neoliberal agenda.

Although Gitlin wasn't at Port Huron in 1962, he soon after became SDS president for a time and engaged in various national actions, including playing the role of "outside agitator" at San Francisco State University under President

Hayakawa. He has become an enfant terrible for the ex–New Left (admittedly, a niche group by now) because he appears to have abandoned leftist politics in favor of a Gulf War–era and September 11 flag waving and because he denounced the Weather Underground in a recent film on the subject. He and his wife hung a flag on the balcony of their Manhattan apartment soon after September 11, expressing solidarity not with the abstract idea of "nation" but with the survivors and victims of that dreadful event. He has recently issued a book, *The Intellectuals and the Flag* (2006a), in which he pursues these ideas. And his *Letters to a Young Activist* (2003a) reinforces the impression that Gitlin has given up the New Left ghost. In interviews I conducted with him, Gitlin told me that he was no longer New Left in his identity because a New Left identity is a period piece of a much earlier and different time. He said that such an identity would have little meaning today, the crux of my disagreement with him (although we share many things, too, that loosely fall under the rubric of progressive politics).

I sympathize with his forward-looking stance (indeed, I treat these issues of the legacy of the 1960s in my 2009 book *The Sixties at 40*). However, I don't think that "the kids are alright" in the sense that they are politically engaged. For the most part, the kids are silent, invisible, absent from the public sphere except as targets of advertising. And it is disorienting to see advertisements for Apple in which characters such as Bob Dylan play a major role. The Jefferson Airplane's song "Revolution" was positioned in an ad for a brokerage house. And John Mellencamp, a 1970s radical of sorts, is now positioned as musical spokesperson for Chevy, deploying a cynical multicultural (better, postmodern) patriotism to capture every conceivable segment of the market for pickup trucks.

And yet I think that Gitlin, in his verve to get on with things (he told me before 2004 that his number one goal was to block the reelection of George W. Bush), misses the evidence of *decline*—of the American Left, of discourse, of books, of young people's worlds and lives. The Frankfurt School brilliantly predicted that progress would be enmeshed with regression, a tendency they trace back to Homeric Greece. They find ample evidence of this in the Holocaust, which was followed by a friendlier (until Homeland Security) fascism in which everybody's lives would be "totally administered," the Frankfurt equivalent of postmodern theory's "disciplining." (The basic difference between German critical theory and French postmodern theory is that the French identify a microphysics of power [Foucault 1977] operating in everyday life that somewhat disables traditional Marxian class analysis. I would respond that Marcuse anticipated all of this in post-1960s writings on the Right's counterrevolution against the sixties, which is still going on.)

Gitlin (1987) argues that the Left had progressive impact on culture during the sixties but that the Right won politics. He means that the Left created a space for civil liberties and progressive sexual politics, but that the much better–organized Right controls statehouses and the welfare state. As of this writing, they hold the House of Representatives and Wall Street, but not the Senate or Oval Office. In a terrific line, he concludes that the Right won the White House while the Left won the English Department! The evangelical Right has forced the Democratic Party to move so far right (e.g., the Clintons) that there is no real choice left for voters, leading many pro-Democrat voters, such as the poor and minorities, to stay home in disgust. I follow through with this argument in my "Beyond Beltway and Bible Belt: Re-imagining the Democratic Party and the American Left" (Agger 2005).

We both agree that the Right has been more tactically astute since the sixties. But one of the Right's strategies has been to demonize the sixties as a harbinger of gay marriage and affirmative action (see D'Souza's [2007] breathtaking book on the putative connection between the campus and cultural Left and September 11 terrorism). The 1960s have become a contested terrain; at stake is the legacy of sixties activism and SDS participatory democracy for kids and adults today. In contrast (with Gitlin), Hayden, Flacks, and Rudd, all of whom I interviewed, identify themselves as New Leftists in their core sensibilities and they all view the 1960s as the best of times, especially in the sense that they formed a large generation of freethinkers and progressives who continue to stand against the hard Right, which has inherited the Earth, or so it seems. They are "homegrown democrats" (Keillor 2004). As it turns out, the dialectic was our friend, Bush and his evangelical foot soldiers "contradicting" themselves in time for the 2008 election and restoring a Democratic administration that has just been reelected for a second term. However, one hopes that the Dems are worth restoring, given their slide toward neoliberalism, toward support of the death penalty, and, perhaps most incredibly, mainstream Dems' partisanship of the war in Iraq and now Afghanistan. The Right's vigorous resistance to a reelected Obama in 2012 may have the effect of radicalizing him, persuading him that a transformational politics, above class struggle and partisanship, is impossible and suggesting that sixties-like coalition building among the young, minorities, and women is preferable.

One of the most important ideas of the 1960s, emanating from *The Port Huron Statement* that itself devolved from a few young Michiganders' involvements with the emerging civil rights movement, was the early SDS's insistence that change worth bringing about must pass through everyday life and affect the self. Hayden's "participatory democracy" was a radical politics of subjectivity and intersubjectivity that connected self and community, social movement, nation, global humanity, and nature. The *Port Huron* campers intended for change to change the self, and for the self to change the world—civil rights workers, antiwar activists, even hippies.

The problem today is that the kids don't protest militarized nation-building or anything else because they are so preoccupied with ephemeral entertainment and commercial cultures that they don't rally, organize, mobilize. As well, they are worried about schooling, jobs, their futures. They stay home, armed with technological prostheses that apparently allow them to be anytime/anywhere. This isn't a real community; it is merely connection. And their anxieties about themselves don't translate into activism in the way they did for youngsters during the sixties, even those who weren't eligible for the draft. In saying this, I am not pretending smug superiority (because we boomers are also self-involved in large part because the New Left politics of subjectivity—think of feminism—requires change to begin at home and return to the home). I am trying to figure out why millions of Americans do no more than buy anti-GOP bumper stickers. An emblem of these pre-political times was a little clock available for ten dollars or so that electronically counted down the number of days until the end of Bush's dreadful second term. Cute, but it doesn't anticipate the very real possibility that Bush would be followed into office by another neoliberal, xenophobic, make-believe evangelical who demonizes the sixties as a rationale for wrecking the world like Sarah Palin. I hope profits from the sale of the Bush countdown clock went to worthy progressive causes.

Obama won the White House in 2008, and kept it in 2012, largely because minorities and women turned out to vote in massive numbers. Young people, too, supported Obama, but their turnout was not significantly greater than in other recent presidential elections. The young were tempted by Obama's promise of a transformational politics that transcended ideological bickering, but he did not keep their attention because the Right blocked—and blocks—his legislative agenda. As of this writing, the majority of Americans, especially the young, support gun control, but, even after Newtown in December 2012, the NRA, GOP, and Tea Party obstruct even modest reforms. These defeats (within an overall reconfiguring of a Democratic majority composed of minorities and women) further turn kids off politics. Young people were politicized during the sixties by a combination of televised Klan violence in the South and the Vietnam-era draft.

The New Reading and Writing and Changing Activism

Kids, even parents, don't read tomes or treatises. Arguably, such tomes and treatises don't exist, having been replaced by fast cultural product—the staples of trade publishing, movie studios, television networks, journalism, even academia (as Jacoby argues). Books can no longer change the world. They don't matter in the way they have since the first printing presses. The UC–Santa Barbara sociologist Richard Flacks (2005), a founder of the SDS and major coauthor of the 1962 *Port Huron Statement*, conducted a study of California state system college students and found that few of them read newspapers. Most obtain their news, if they get news at all, from the Internet. Internet news is often superficial and episodic. I'd generalize this and guess that few children of baby boomers (born between about 1947 and 1960) read books or even magazines. My college students usually sell their class books back to the campus bookstore after the semester is over, or they just rent them in the first place.

Young people today did not grow up in a bookish culture. The surround-sound media culture never gave them a chance. Their options are television, radio, the Internet, cell phones, and iPods. Baby boomers had black-and-white television (three channels) and a transistor radio. I inhaled my San Francisco Giants games with the radio glued to my ear. What I learned I mainly learned from books and magazines. I started a small newspaper when I was about ten. That was my literary outlet for a while!

We boomers had real choices. Access to popular culture was not 24/7; it was not inescapable, and people still went to libraries and read long, slow books, both fiction and nonfiction. Books then could change consciousness as could journalism, such as reportage of the Vietnam War. We had few electronic options, fewer technological prostheses connecting us, "disciplining" us. Friedan's *Feminine Mystique* (2001 [1963]) and Cleaver's *Soul on Ice* (1968) started arguments and even social movements.

However, there may be a literary impulse—to write, read, create, cajole—that a totally administered cybercapitalism cannot contain or divert harmlessly. Anyone watching preschoolers and even early elementary kids fill a blank page with images and text realizes that the literary impulse is irrepressible. Kids may not read in the traditional sense, but they lead active literary lives. There is a dialectic at play here, as there always is in an unstable capitalism. Although kids don't inhabit a pulpy world of tracts and treatises—nor write them—they nonetheless lead active literary lifestyles. That should give us hope, if not comfort. They wordsmith emails, text messages, and tweets, quickly crafting them using shorthand, slang, and emoticons—LOL. A posttextual world of images, videos, and television is not entirely devoid of literary forms and institutions. Kids do write and read, although not *The Port Huron Statement* or *Autobiography of Malcolm X* (1999). They want to lead textual lives in order to connect, framing their identities, finding friends, and establishing romantic relationships. Kids may have abandoned traditional pulp media, but they have embraced social-media postings, texts, blogs, and emails as alternatives. They consult a variety of sources and visit a range of sites, such as Jezebel.com and Rookiemag.com. These kids, perhaps somewhat offbeat and intellectual, participate in a cyber-sphere in which people discuss ideas, issues, fads, foibles. Some of my college students are regular visitors at huffingtonpost.com. As I noted in the Preface, 40 percent of what kids read is found on various social-media and websites that, taken together, constitute a nontraditional body of knowledge and, potentially, a canvas for their own creations.

As I explored in the previous chapter, kids' adoption of new literary life forms has three implications. (1) They seek uncoerced instantaneity, the possibility of making connections without necessarily having to respond. (2) They seek the ability to manage impressions, composing new identities on the screen that are plastic and can be changed. Finally, (3) electronic discourse, composed hastily, is quite depthless, given its brevity and its topical nature. Although it is possible to write lengthy tomes using electronic means, this is an age of instantaneity in which people seek connection now, in the moment, even as they want the ability to filter. It is not the electronic nature of the discourse that destines discourse to be depthless but the way in which fast discourse, made possible by these new electronic means, is rather non-analytical but self-expressive. Texts and tweets are usually about "me" and not about the larger world. As Gitlin indicates, the Internet can be used to organize as well as analyze. I choose to view electronic media as having dialectical potential; it can become more than it is, achieving depth, relevance, and soulfulness. Yes, the majority of tweets and emails are about mundane things, but so is the majority of conversation. Most communication, in general, is topical and mundane. But along with these mundane digital communications are a treasure trove of tweets and Facebook statuses that are political and cultural, that deal with ideas. This is perhaps a new genre of political communication, one that is made up of pithy arguments that accumulate over time into a substantial political statement, albeit one that is serial rather than unitary. I am hopeful within pessimism!

Connection without Community, Anxiety without Activism

Kids seek connection but not community; they are anxious but avoid activism. We need to guide them toward more public lives. Adults, by default, run the world. But we do this in ways that allow young people to engage in self-privatization and then we turn around and blame them for self-absorption. We abandon the young because we are too consumed by our own lives. And we, those among us who profit from fast capitalism, allow them to purchase means of electronic connection that suck up their time and turn them away from reading and writing. And adults are enmeshed in these same circuitries of shopping, diversion, false community. None of us seems to have many options.

When I talk about kids and young people, I am of course identifying that treasured marketing demographic between eighteen and twenty-nine. But I am also including younger teenagers and even thirty- and forty-somethings. Perhaps the most relevant boundary is between people who grew up with the Internet, social media, and texting and those of us (baby boomers) born between about 1947 and 1960. The height of the baby boom was 1957, and I was born in 1952. I am also generally talking about young people who are still in school and who have trouble with term papers but write with versatility in nontraditional, electronic ways. This book is about literacy—literacies, really, because there are several legitimate types of reading, writing, creation. Perhaps we always project what we learned and what we do, reflecting inherent generational bias. But these biases need to be tempered by a real reckoning with the literary practices of our kids and students. E-sociologies are ethnographies of literary practices that, like all good anthropological field work, inculcate tolerance and enhance cosmopolitanism.

What can the elders do to enable their children—children of the children, those who were young during the sixties and put their lives on the line—to read the paper, empathize, and organize? A place to begin, but not the only place, is to teach and write about the 1960s in ways that don't sermonize but explicate the relevance of those times for these times. It doesn't take a genius to discern that the FBI's COINTELPRO program is now Homeland Security, Vietnam is Iraq, blacks are gays and lesbians (and blacks and browns, too), Nixon is Bush. The parallels are obvious, but only if kids understand enough about America forty years ago to make the connections. I show movies to my students about Kent State, New Left protest (including one on the Weathermen), and antiwar Vietnam vets in order to bring my visually oriented students up to speed. To be sure, that would seem to defeat the purpose of reviving paper as a medium of enlightenment and mobilization. But I also assign books on the sixties such as Jim Miller's *Democracy Is in the Streets* (1987), Sale's book on SDS (1973), and Hayden's *Reunion: A Memoir* (1988) for students who want to go deeper. If I started with the books, few would read them. The movies present some of the same footage that astonished and outraged us forty years ago, but in a culture in which almost everything else was learned from print.

So teaching the 1960s is one way to begin. This empowers students who learn that Hayden and the other young Michiganders who convened at Port Huron did nothing less than change the world with a mimeographed document drafted by a college student. My students above all feel disempowered by the world that bears down heavily on them. They struggle to pay their bills and their tuition, to find part-time jobs and then establish careers, to forge ahead by establishing families. The seemingly limitless frontier opened by Kennedy's New Frontier is a distant memory, if it is a memory at all, to students who have never heard of Watergate, Monicagate, and Iran-Contra. As of this writing, Jeb Bush is considering a presidential run in 2016, making the name Bush seem dynastic. Our times feel partly like the 1950s in that many are quiescent, and partly like the 1970s when the hard Right consolidated their victories over the New Left and minorities. Except now there are more vehicles for escape and avoidance than were available during the 1970s, when the world was still partly available through pulp. Before the Internet, there were boundaries, which allowed people to retreat and reflect.

Adults can do something else that would be helpful. They can model activist behavior by being activists themselves. Why should the burden be on college students and other youngsters already overloaded with expectations to produce and perform? Hayden and the happy campers were an anomaly; college kids starting a revolution by writing a document happens once in a blue moon, at best. Indeed, we graying boomers are well positioned to take the lead in antiwar protests and other acts of civil disobedience simply because we have already done it. We understand what it is like to march on a lonely picket line being heckled by hard hats, to go to jail for our conscience, to deal with family members and relatives who disapprove of us, even to leave the country.

Finally, and perhaps most to the point, we can reintroduce our children and students to a world full of pulp. We can encourage them to read, and not just emails and texts. We can encourage them to write, not just the flotsam and jetsam of episodic cable traffic but journals, articles, blogs, publications. We can take them to libraries. We can read and write in their presence. We can use visual tools and techniques, such as the 1970s documentaries mentioned earlier, and enrich them with further reading. We can refuse to use short answer "objective" tests in our classes but ask them to write—not just term papers, easily forgotten once written and easily purloined from the web—but journals that record students' coming to grips with the implications of their readings for their own lives. When I did that in my classes, many students began to view their writing as an opportunity to pour forth their souls, not as a punishment to be graded in the dreary production process of human capital (credentials). We teach them that writing need not be purely objective but can express passion and celebrate experience, discarding a positivist model of scholarship.

I am not the first to lament the attenuation of considered reading and writing. Longreads.com is a site that celebrates the essay, which is Montaigne's word for self-creating prose; essay in French means "to attempt." Of course, blogs are attempts in this sense. Electronic communication tempts people to go short, thus avoiding complexity and depth; think of Twitter. But the Internet always creates a publishing revolution as everyone can join the conversation, without the gatekeeping of cultural elites who keep outsiders out. But gatekeepers also enforce standards, as non-postmodern academics remind us. The problem is that standards change; they are always already fluid, and they depend on perspective. Anyone who does university-based writing knows that peer-reviewed scholarship is fraught with contention and conflict as reviewers often disagree about merit.

My argument that the decline of pulpish knowing, reading, writing, agitating, and organizing is responsible for the political impotence of young and older Americans today can seem rather like the argument that academic Leftists cause terrorism. However, I am not saying that the decline of the book *causes* political apathy but rather that political apathy and the decline of books come from the same source, and they reinforce each other. That source is none other than a postmodern fast capitalism in which old-fashioned argument is concealed in, and congeals as, what I call secret writing. And these detextualized writings ooze out of the covers of books and bleed into the world, commanding lives in secret, subliminal ways. The "books" today are text messages, advertisements, gendered body positionings, even—no, especially—positivist social science, which reflects a frozen world in order to freeze it. This representational version of science quietly understands that the world is in fact not frozen but molten, much as Marx said it was. Positivist versions of the world seek to make up for the deficit—call it agency—that prevents the social world from proceeding according to putative laws of social motion, for Parsons patriarchy, the market economy, civil religion, and so on.

And so we must be quite clear what has happened to books and writings, such as *The Communist Manifesto, The Other America,* and *The Port Huron Statement* of the SDS. These are contained in libraries that, even on university campuses, have turned into lounges and cafes. Books are now to be found everywhere but in the library, online and in texted messages, on television and in the movies. They enter the world silently, much as all fetishized commodities do; these commodities conceal the deep struggle and inequities that brought them into the world. By commodity fetishism—the essential feature of fast capitalism—Marx meant the concealment of the labor processes and struggles that underlie the wage relationship. Bourgeois economic theory purports to represent a world operating smoothly according to market laws.

Kids today do not dig underneath this sedimented world to discover the authorial pulse underneath these commodified and dispersed texts, even their own. Kids busy texting each other don't imagine themselves to be their generation's Tom Hayden, Graham Greene, Ralph Ellison, or Ralph Nader. They plug in and recharge their batteries, taking for granted the burning world around them and even their own misery. A few driven young entrepreneurs found tech companies and become dot-com millionaires, but this is capitalist business as usual by another name. It is highly debatable that Apple and Facebook will leave an imprint in the long run.

Would *The Port Huron Statement* have resonated today if it were introduced as a niche Web page? Although written to be read by students as well as theorists, *The Port Huron Statement* was lengthy and required more than an hour's work. It was passed around and pored over. It was discussed in places that didn't have an SDS chapter. If it were produced on a website today, it would certainly be more accessible, but it might be more easily ignored as well. Perhaps the very difficulty of producing and distributing writing in the 1960s and the relative infrequency of publication meant that those writings that were published garnered more attention. It is easy for posts today to get lost in the vast Sargasso Sea of the Internet. It's also possible that fast capitalism and its digital formats mean that writers have lost their distance from the world and thus their ability to understand the world critically and to think it otherwise, to utopianize and to energize. We must find ways to reach young people, to encourage them to find their political voice, create means of writing and reading the treatises that would change things, of the kind that we used to read before the 1960s got sidetracked by the Right's counterrevolution, which proceeds apace.

The Internet is capable of supporting a political culture. One can post lengthy, thoughtful work in e-books, blogs, and online newspapers. But digital technologies allow and cater to instantaneizing tendencies, ultimately, by its ease, encouraging the compression of discourse into temporally tiny units and exchanges. LOL saves keystrokes and cell charges, but it does not rival thicker description of what the author is feeling—laughing out loud, perhaps reveling in irony, engaging in sarcasm, who knows. People's, especially kids', attention spans have shrunk in our ADD-icted culture. We are impatient, we have been *made* impatient, by a restless fast capitalism that requires quick as well as ceaseless production, circulation, and consumption of commodities, including cultural ones. We are overwhelmed with electronic access, as cable and satellite television, to cite only one plausible example, offers too many channels for any busy viewer.

Discourse has declined because it has been accelerated, the argument I made in *Fast Capitalism* (1989a). That book was written when the Internet was just taking off, and I could not have imagined the extent to which the Internet would quicken the pace of discourse. Even in my sequel, *Speeding Up Fast Capitalism* (2004a), I did not deal sufficiently with the attention-depriving tendencies, especially for kids, of texting, messaging, and surfing. Perhaps having teenagers gives me access to a pulpless kids' world in which few care about the war or anything else beyond their immediate lives. In Texas, where we live, the fast cultures of kids are mediated by evangelical Christianity, which adds yet another layer (with due apologies to postmodernists and liberals!) of false consciousness—a systematic, engendered inability to see the world for what it is.

As I finish this chapter, destined for academic and other well-educated readers, I could also post it as a blog entry or quick e-article in order to produce a few more ripples. Pulp needs help from electronic dissemination, which can, after all, be downloaded and printed out. I am not arguing for reboundarying the pulp world, decontaminating it by ensuring that it cannot go electronic and become instantaneized and hence depoliticized, but for versions of analysis and argument (issuing in activism) that reach readers, globally, who don't have access to university libraries or independent bookstores. Perhaps in this way—again dialectically—pulp can use depoliticizing media against themselves, ever the agenda of a 1960s politics of subjectivity and intersubjectivity that burrows from within.

We know from previous eras that electronic media—like television in the 1960s with its coverage of everything from the JFK assassination to civil rights and Vietnam—can produce densely analytical discourses that deserve and demand careful consideration when literate writers and readers are situated already in a public culture of ideas. Whether this public culture of ideas can be created using the Internet and other instantaneizing technologies is an open question.

In the next chapter, I develop an "e-sociology," as I call it, appropriate to understanding the life forms and discursive practices of youngsters. The boundary between the Gutenberg and post-Gutenberg generations is permeable. Adults, too, email, blog, text, and tweet, which, along with message boards, ushers in an Age of Opinion in which everyone has instantaneous expertise. Texting has utopian and dystopian implications, opening a dialogical world or—and—rendering us alone together (Turkle 2011).

CHAPTER 5
E-SOCIOLOGIES OF VIRTUAL SELVES

While young people don't read much in the way of traditional pulp vehicles such as books and newspapers, they do pursue active nontraditional literary lives—primarily lives on the screen. They may not do their homework, but they furiously compose text messages, tweets, Facebook, and blog postings. They email. They seek uncoerced instantaneity, the ability to manage impressions, and a hurried discourse that pulp generations might view as shallow. Above all, kids seek connection, if not full-blown community from which political and social movements spring.

The E-Sociology of Children's Communication

Traditional sociology has rarely theorized children and has only recently begun to theorize the virtual worlds of Inter(net)actions. As well, there is a bias that a literate culture is necessarily based on pulp media. An "e-sociology" might read electronically mediated literary practices as valid and valuable in their own right. This sociology might also examine the issue of whether young people, plugged in and virtually connected, can be politicized, given their apparent political lassitude. Perhaps texting and blogging are signs of resistance and not simply acquiescence to a media culture in which instantaneity thwarts sustained reflection and critique. Or perhaps all of this furious literary work in pursuit of connection renders impermeable the boundary between adults' and children's worlds, making it even more difficult for adult sociologists to figure out what is going on.

Children are missing from sociology and social theory; they are the last minority group. Family sociologists treat issues of development and adolescence, especially in a family systems context. But there is little serious work devoted to childhood and adolescence as subordinated moments and practices. Few speak for children, who, in fast capitalism, are buffeted by social forces—the Internet, violent media, pornography, dating, drugs, and alcohol—nearly out of their control. Children have become virtual, in both senses of the word: They live their lives on the screen, and they have been nearly effaced by adults who burden them with impossible expectations. E-sociologies are sociologies that read and learn from the secret writing and living of youngsters—members of what I am calling the pulpless generation. These sociologies depart from Parsons, who viewed socialization as the ingestion of culture and norms. They also depart from more conflict-oriented sociologists who focus on adult labor forces and possibly children's roles within them. These sociologists may tend to portray kids as passive receptacles, not also as agents. E-sociology rejects a view of children as

- Parsonian role sets who internalize norms
- Passive consumers buffeted by the culture industry
- Pre-adults who grow naturally into full citizens
- Potential members of deviant subcultures

Instead, e-sociology views children as *agential,* if also as *accelerated,* both authors and victims of their own lives. Its data are the tracings of social-networks found beneath the manifest data of adult life. Social media like Twitter allow people to track the text messaging of other users; they constitute electronic traces of a life lived on the screen and on the phone, which, interestingly, is used less for talk than as a textual vehicle. I ask my kids and their friends why they text and just don't talk. They look at me as if I come from another planet or another time. They know why they text: They want to author their lives unconstrained by adults, and texting is a form of writing. Today, phones and computers are vehicles of self-creation and communication. Texts and emails are composed quickly, even hastily, not for posterity. Shorthand is used for reasons of economy and also to defy the adult censors. In this sense, perhaps these are less mediated versions of the self, who pours her heart into these connections, which constitute social relations. There is a prejudice that real writing is done slowly, from a distance, and endlessly revised. But writing can be done nearly at the speed of light, where kids write secretly in order to build a community of non-adults who share antipathy toward adult authority. Messaging can be viewed as a political act where it is intended to create a polity—perhaps groups of friends who share a taste in music and despise a certain teacher, who is described as lame or worse. These youthful polities don't necessarily participate in the political process or even vote. And yet, for the first time since about 1968, when youth gravitated to two Democratic presidential candidates, Eugene McCarthy and Bobby Kennedy, kids are excited by a national politician in Barack Obama, who promises a transformational politics.

Viewing children as agents is akin to viewing minority group members as full citizens. It liberates them, giving them rights and the right to be seen, viewed, heard. It is different in that children are not yet fully formed (if indeed even adults are). My son sometimes says (and writes, texts, messages) things beyond his age, while other times he seems child-like. This raises questions about age-related boundaries like the legal drinking age, voting age, military age, driving age, retirement age. These boundaries are highly problematic when treated rigidly as developmental stages. People age and mature in various ways, and for most these processes are not linear. The Victorian concept of the child, although intuitively appealing especially when compared with the harsh medieval view that children are simply adult-like members of a vast work force, deconstructs when examined closely. When does the infant become a child? The child an adolescent? The teenager an adult? Few would defend the eighteenth-century view that seven-year-olds should be working in factories. But should sixteen-year-olds work long hours or night shifts? Should eighteen-year-olds go to war? Should vibrant people in their mid-sixties be forced to retire?

One way to investigate these issues is to collect data from the people—young people—themselves, doing ethnographies and discourse analyses of their wants, needs, hopes, ideals, values, transgressions. Just as "self" and "identity" are acquired over time, if they are ever fully acquired, so we must read the traces of self-expression to chart our children's development—and hence their agential status. E-sociologies would collect data from their communications, bloggings, postings, and texts. These data would disclose the self as it emerges even from early childhood. These traces are the outlines of identity, of values, passion, even politics. They may be alarming to parents and other adults. Some may be inflated or even faked, as kids explore the plasticity of the virtual self. What Turkle (1995) calls life on the screen affords opportunities for self-transformation and self-positioning, but it also provides adults (or those of us who pass for grownups) information about the growth of the young self. Of particular interest is the extent to which they recognize themselves as agents, people who are at least somewhat in control of themselves, their bodies, their education, their work, their future.

The primary experience of childhood today, in our peculiar era when we disempower children but accelerate and overburden their lives, is of being trapped, even imprisoned. They move from one subordinate status to another, from K–12 student to part-time worker to college student to adult apprentice in the nine-to-five rat race. Homework is assigned them and they are confined in prison-like school buildings with scant time to exercise and recreate. They are dressed in school uniforms or standardized dress. They are profiled by police, parents, and teachers as likely criminals or at least deviants.

Identifying the Rules of Cyberdiscourse

Wittgenstein (1958) coined the term "language game" to refer to the unwritten rules by which discourse proceeds. For example, we all understand that on a first date, you don't ask your partner to marry you. We understand that, when standing in line at the grocery store, you shouldn't ask your cashier to have sex (although who knows these days!). Sociology has its own language game, in which it's understood how, as a discipline, it should proceed from literature review to methods to findings—rules that almost always impose narrowness on the research and results.

Kids' cyberdiscourse can be understood in terms of language games. In the following sections, we will parse the rules of the language game for cell phones, texts, instant messaging, social-networking, and blogging and see what they can reveal about the agential adolescents doing the conversing.

Cells

We have all been with a friend who hears his cell ringing, takes a quick look down, and then returns his cell to his pocket. This marks one of the central rules of the language game of cell phones: You can look away. Ignoring an incoming call is a form of what earlier I called mediation. You pick and choose to whom, when, and where you want to talk. Talking can be inserted into everyday life without missing a beat. You are always available, but you can never be coerced into feedback. My students check their phones after class to see who called. Cells are lifelines, the muscular connections of the children's prison community. But cell phone calls also require synchronous, real-time interaction. You cannot defer your answer to "Want to go to the movies?" If you do choose to engage, that engagement is immediate; there is no time to think about a reply, to consider before committing to a response.

Texting

Texting takes place on cell phones but has distinctly different rules from phone calls. Texting shares the advantage with cell calls that you can turn away. But, unlike phone calls, you can choose to engage in a less immediate manner, pausing, considering, typing, and then deleting. The fact that texting involves typing also means that texting requires more attention. You have to engage the technology with both hands, keying the message. Texts on certain phones allow only a limited number of keystrokes and so there is a lot of back and forth; planning a date may take ten exchanges. Also, to suit its peculiar literary economy, the text will probably be delivered in code, with acronyms abbreviating spelled-out terms. This tactile engagement is straightforward in most contexts but can become lethal when combined with other complex tasks like driving. The major advantages of texting for young people is less surveillance by adults and the ability to mediate the interaction. One can avoid direct questions.

Instant Messaging (IM)

IM involves sending messages over a social-network program that allows real-time communication. It allows longer messages than texting, but, still, full sentences are rarely used, with users preferring acronyms and abbreviations. Messaging is most satisfying when one has a full keyboard, for example using a laptop or tablet, and can devote sustained time and effort to full exchange. But there is more interactional coercion in play: It is more difficult to be evasive than with texting unless one is totally rude and just signs off. And the rate of play is greater than with texting, which involves smaller keys and a more strenuous physical protocol.

Social Networking

Social-networking refers to the hundreds of websites and programs that allow users to identify friends (or make new ones) and communicate easily. Some are general (like Facebook) and some are specialized, focusing on a particular interest. You can email and message through social-networks, and you can exhibit yourself, creating a virtual personal web page, with images, vital statistics such as relationship status and personal preferences in bands, television, food, culture heroes. These can be open for anyone's reading or restricted to your friends. Social-networks make it possible for large numbers of friends to interact together, as a community, rather than one-to-one as they do with texting and messaging. The opportunity for frequent status updates means that kids are essentially journaling their lives.

The downside (at least for kids) is that social-network sites are often more transparent. Many parents insist that they be friended by their children so that they can monitor their activity. But even when this isn't the case, the relatively public nature of Facebook means that information and images tend to get shared and circulated. Parents learn who their kids hang with, and many a parent has gagged when they discovered via Facebook that their kids have used drugs or had sex.

Blogging

More literary and older kids blog; they put up web pages that double as full-fledged journals, even inviting responses from others. Blogs can have a commercial use when one provides links to one's paid work, or when the paid work is virtual work. These are feasts of self-expression, documenting one's inner and outer journeys. They take time to write and to read. Do blogs get read? My hunch is that more people read blogs than have read all of my books combined! Blogging is perhaps the one area of digital communication where kids *can* imagine themselves as the next Tom Hayden or Ralph Ellison or Ralph Nader.

All of these digital communication modalities interpenetrate and overlap. Most kids who text also tweet, use social networks, and message. There are many conversations going on, even at the same time, with different informal rules of the game. The quick text or tweet tends to be favored over voice-to-voice full disclosure. Kids can be insecure about their identities, and texting allows time for a considered reaction that voice-to-voice doesn't permit. And, of course, texting allows them to hide from adults. Kids' cybercommunication can be all-consuming. They spend many hours of their down time (and even some of their up time, during the school day) making connections, chatting, expressing themselves. They write thousands of words a day, even if these don't count toward their grade in English. This may be seen as a kind of prison talk, a venerable vehicle of public discourse composed while in private. Antonio Gramsci composed notebooks in prison, and George Jackson did the same with *Soledad Brother* (1994). Some posttextual teachers count and encourage blogging, as well they should. But this is relatively rare, especially where teachers are hamstrung by a rigid curriculum anchored in yesteryear, before electronic communication.

Cyberdiscourse in the Context of Postmodernism

Emily Gould, a noted young blogger, cautions that in blogging people may "overshare," making the private too public (2008). She argues that people want to blog in order to leave some trace of their lives, to reassure themselves and others that they matter. This may be an important part of identity formation for young people, at risk of losing themselves to the world. Although we elders may doubt the significance of leaving electronic traces of the self—who I'm dating, what music I like—the significance lies perhaps in the fact that these traces are signifiers of the self, proof that one matters. She has done some oversharing herself, in her memoir (2010). But haven't we all? Especially when we accept the postmodern idea that all writing is autobiography (as well as autobibliography).

This may be a boundary between the modern and the postmodern: We now write for its own sake, for self-expression, and not necessarily to be read. Journaling keeps us sane and together. Writing integrates identity. Using Twitter, we can chronicle our day, adding narrative structure to our lives. This is a kind of literary exhibitionism, also displayed when people check in with each other via their cell phones, either voice to voice or by texting. "Wassup?" "Watcha doin?" These electronic connections may be the sinews of a postmodern society, replacing physically shared space and face-to-face interaction as the ground of identity and community.

Virtual selves are no less real than embodied ones. Or is it a new kind of reality, a pixelated one? We come to experience others through the screen, changing the nature of talking and writing. PowerPoint when used in teaching has much the same effect. The teacher reads the screen to the class, who copy it dutifully. In effect, like medieval monks, they are copying the textbook, which is projected above the teacher's lectern. One might say that PowerPoint is reinforcing, the screen redoubling and magnifying the message of the printed page. Watching teaching or public speaking with PowerPoint is like watching a prayer session; students and teacher worship the text projected above their heads.

Elders like me have an easy time denigrating these devotional activities as reifying—literally, "making real." But Derrida is correct that oral speech, conveying presence before the Truth, is not necessarily more truthful than the text, less free of error. He denounces the "metaphysics of presence" on which modernism and realism have been built. For someone like me who laments the decline of books, postmodern textuality might be a revivification of texts, whether via PowerPoint, text messaging, or other sorts of posting.

I don't really believe that. Watching teaching via PowerPoint strikes me as profoundly lazy, the instructor simply projecting the text and then reciting it, even organizing student note taking with "take-away points." I've often wondered about the very act of summarizing, of bulleting important points (as I have already done in this chapter). Are the other points somehow less relevant? If so, why bother with them? Bulleting hierarchizes—that which is to be taken away and that which is to be left behind. The non-bulleted may be the most interesting things, the more scandalous knowledge. Derrida was obsessed with margins and marginality, fearing that the center will not hold.

Walter Benjamin (1969) of the Frankfurt School, like French postmodernists, would have been a terrific e-sociologist of virtual selves. He felt that general knowledge could be refracted, prism-like, through various sundry particulars—perhaps scribbling on an envelope or instant-message exchanges. He valued the particular because, in line with his Frankfurt colleagues (all German Jewish intellectuals), he worried that the particular would be exterminated under the totalitarianism of the Enlightenment in which only quantifiable, "positive" knowledge would count. Benjamin lost his life at the hands of the fascists in spite of it all.

And so virtual selves cling to selfhood in an overwhelming age. The self is overwhelmed by information, expectations, busy work, globality, stimulation, sexuality, discourses. Laptop and cell phone are grounding when there is no ground, or, better, when every ground shifts and gives way. In my childhood, the ground was firm; we had neighborhood, friends, school, parents, even as, by the mid-sixties, our world was ablaze with the conflicts and passions of civil rights and antiwar movements. Even then we were on stable ground; the sixties were pre-postmodern. We were the Cleaver boys, simply radicalized by tumultuous events. The postmodern only emerged later, with right-wing extremism, counterrevolution, globalization, perestroika, and the Internet. Arguably, the space race started it all, with the miniaturization of computing. And the space race was triggered by the Cold War and Sputnik, the postmodern emerging dialectically from the seemingly firm ground of the modern.

So perhaps there is no firm ground, or the ground is firm everywhere, not only in places on the official map. We are always at home wherever we are, including in the virtual world. Home to me when I grew up in the stable college town of Eugene, Oregon, is perhaps unavailable to kids today whose parents both work and they return home alone and seek electronic connections. This sounds like a lament for yesteryear, and perhaps it is. But our children have opportunities we never did for connection and community albeit of a kind not immediately recognizable to many elders who still view identity in terms of a firm mooring in place and space. Again, the issues of the modern and postmodern and of boundaries.

Derrida in appreciating marginality and Benjamin in valuing the fragment provide coordinates for e-sociology, a sociology of traces, of everyday life, of discourses. What matters to modernist sociologists, and modernist selves, may not be of moment for our children, who care less about paradigms than about perspectives. I am tempted to condemn the postmodern as an artifact of the technology and communications companies and of capitalism generally. But I catch myself because—again the issue of boundaries—I am convinced that the modern has not segued completely into the postmodern but rather that they overlap and blend. We still have books, arguments, neighborhoods, paradigms, politics, even revolutions. But we also have secret writing, perspective, cyberspace, and both the pre-political and post-political. We are still, in the words of the sixties bards Simon and Garfunkel, looking for America, but we realize that there are no prepositions such as "for." We are already there, home, but we are running away or running toward, failing to recognize that we are all moving forward together. Perhaps runners understand this better than philosophers and certainly better than scientists. After a while, and many miles, they realize that they are running in place. They are still stuck to the ground, that no matter how fast they run, they do not take flight. Gravity grounds them, and even running across the country takes them nowhere they haven't been before.

E-Sociology as Part of the Quest for Children's Rights

Later, I will tell of a young man named Shapiro, who ran across America in the early 1980s. He was profoundly changed by his three-thousand-mile journey.

He discovered that the effort expanded his perspective on himself; he could do something he hadn't thought he could. We don't have to run to learn this. We can all inhabit the places of single-minded effort, whether we play the guitar or write or run. Our children have boundless energy, much of which they devote to their secret writing. But they would not feel like prisoners if elders listened and appreciated. I am close to believing that the young are the new proletariat, which we certainly believed during the sixties. The SDS talked of a children's revolution, a revolution of students. We viewed ourselves as an oppressed class, ignored by elders, especially by our teachers, who did not serve our interest in "relevance." Perhaps this is the way that the generations always relate to each other. I would like to believe that the elders and the young could achieve solidarity across their profound boundaries.

We ignore the young but we all turn old. Adolescence is a temporary station (fortunately!). However, childhood does not disappear; children age but they are replaced. The civil rights movements of the 1960s bore the environmental movement, with the first Earth Day occurring in 1970, as one decade bled into the next. And the appreciation of nature as a victim of civilization that could be protected led to animal rights, showing that civil rights, broadly understood, provides protection for all sorts of "subjects," certainly, I would argue, including the young.

Sociology tracks the broadening of rights from antiquity through the Enlightenment and through to the present. Initially, Plato, Augustine, Rousseau, Paine, Marx, Weber, and Parsons wrote of the rights of "men," literally male property-owning citizens. Women, children, slaves, and non-Europeans were in the closet, doing important but unvalued work. The Enlightenment offered a broader conception of rights, and then the social movements of the 1960s broadened that broadening even further—even, eventually, including nature and animals. But sociology still largely ignores the children because we tend to ignore the possibility that children, as a class, can be oppressed.

Child abuse has been noticed, analyzed, deplored. But the abuse of children as a group tends to go undetected, their being ignored constituting the problem. Children themselves go undetected. I track children's writing as I develop further an agenda for e-sociologies that establish context for the reading of children's electronic texts (postings, texts, tweets, instant messages, emails). These sociologies inhabit children's everyday lifeworlds as they distinguish between disturbing cries for help by kids and kids' utopian intention to create a world not without adults but without untoward adult authority.

This is to suggest that kids are political and social theorists; they make sense of their worlds, especially as they grapple with the thorny issue of whether adult authority is the same thing as adult authoritarianism. Sociology has tended to ignore kids and their electronic discourse, just as sociology, unless it is Marxist, tends to ignore the boundary between authority and authoritarianism. Perhaps another way of saying this is simply to note that few sociologists have been utopians, except during the exciting years of the 1960s, when they fought alongside of, and educated, civil rights and antiwar activists.

Harold Garfinkel (1967) reacted against Talcott Parsons's leaden 1950s functionalism by arguing for the relevance of everyday life; his ethnomethodological program tracked the practical reasoning of people living their lives without a Parsonian script spelling out their prescribed roles. He made possible a sociology in and of the natural attitude, the everyday orientation of people to the challenges of making a living and raising a family. This sociology is very much in the spirit of the e-sociology of virtual selves. Garfinkel paid attention to discourse, the ways people talk and write. He did not focus on children or teens, but his program is easily adaptable to the concerns of this book. He would have been fascinated by posting, texting, and messaging, as these are everyday ways of conducting life with others.

As well, they are amateur and "indie," neither supervised nor governed by rigid rules. The informal norms governing the use of Facebook and text messaging emerge from the users themselves, many of whom are youngsters. E-sociologies are ethnomethodological, in Garfinkel's sense, because they trace society and social structure from the ground up. Kids improvise the roles that they play; they creatively text, game, download, chat. This is not to deny that there are technological limits to what they can do—and write; think of the limit on characters in text messages on certain phones. Nor are we ignoring the power of advertising, within the culture industry, to create needs and govern taste. Kids are drawn to iPhones and iPods because they have been exposed to marketing, and they mimic other kids who have and use them. And yet once equipped, kids manipulate the technologies of communication and entertainment in ways that aren't taught to them in school or by adults. They make it up as they go, which is what Garfinkel says we all do, all the time.

Garfinkel implies that sociology can be conducted within everyday life; indeed, it is always/already an activity of everyday life, even if the living is being done by college professors with advanced degrees. I have studied the ways in which social scientists write, using insights from French postmodern theory and the

Frankfurt School's critical theory to illuminate the busy scribbling underlying the smooth published journal page. Sociologists have everyday lives of their own from which they can never gain full distance. These lives include making sense of their worlds, which involve a certain scientific sense making. They—we!—make sense of ourselves making sense, as we build academic careers, write and edit ourselves, teach, publish.

This is less an unmasking, observing how the sausage is made, than an empowering of anyone who lives in the world to think, write, and act sociologically. By *sociology* I am referring to writing about the social world that is aware of itself writing and that refuses to purge the authorial self from the page. We are doing sociology when we include ourselves in the stories we tell about the world, our world.

Sociology asks what makes us who we are. Knowing this can help us change ourselves and even the world. E-sociology explores our imprinting by our invasive media cultures, and it uses culture against itself as it composes itself reflexively for all to read. This sociology blogs—another word for, and way of, publishing. Increasingly, academic publishing is being pixelated, as pulp publications go online. Academics may worry about who is vetting all of this posted work; is it refereed, read and evaluated by other experts in the field? Posted work can be screened just as traditional pulp work is screened by knowledgeable experts. But the exciting thing about online writing is how it is virtually (pun intended) impossible for expert gatekeepers to keep amateur and indie writers and creators on the outside looking in. Everyone can compose and post sociology.

The former positivist illusion that sociology was not actually scripted by perspective-ridden writers succumbs to the realization that sociology is just another way of being in the world—an especially reflexive one. By *reflexivity* I am borrowing Gouldner's (1970) and O'Neill's (1972) term for writing that observes itself writing, and living. It is writing that not only does not banish the self from the text but requires texts to be anchored in experience, perspective, passion, politics. As such, a reflexive everyday sociology—both addressing and grounded in everyday life—is a political practice, attaching value to discourses and practices heretofore largely neglected. This is very much the agenda of a dialectical anthropology (Diamond 1974) that engages with peoples and practices typically neglected by official social science, especially where those people and practices are in "primitive" worlds. Diamond aimed to blend primitive and modern in much the way I advocate a "slowmodernity" in the following chapter that blends fast capitalism and earlier "slower" forms emphasizing our metabolism with nature, animals, bodies.

And so e-sociology is a matter of both content and form. It engages with children and teens at the level of their everyday discourse and practice. It puts kids on the agenda, both sociologically and politically. Necessarily, it addresses intergenerational politics between kids and their parents. As well, e-sociology is not only *of* everyday life but also *in* it; it is a mode of being-in-the-world, a life dedicated to writing and reading toward social justice. Here, this writing takes the form of readings of youthful discourse that treat them as symptomatic of alienation and also as a utopian prefiguring as children "do" the world in non-hierarchical and non-authoritarian ways.

The space race required miniature computers, which led to PCs, Macs, and then phones that doubled as computers. In the following chapter, I consider the ubiquity of smartphones and consider the impact on time, textuality, and life itself of a new, postmodern type of time that I dub "iTime."

PART II

TIME REBELS

Chapter 6
iTime

Everyone under a certain age in affluent industrial nations has, or wants, a smartphone—iPhone, BlackBerry, Droid. And that "certain age" is rapidly increasing, as even the elderly become more adept with email and texting. Smartphones (oxymoron? tautology?) change the ways we work, live, sleep, connect, and do community. As ever, our kids are out in front of social change. I consider these changes dialectically—that is, in terms of upsides, downsides, and the potential for further change. My main thesis is that smartphoning creates a kind of "iTime" that challenges the pre-Internet boundaries between public and private, day and night, work and leisure, space and time. Our kids experience this time as normal, while some elders view it as a nightmare of connectivity and accountability. Images of utopia, of the good society, vary, in some measure generationally.

Labor's Internet Commodity Form

Physical labor, for much of the workforce, has been replaced by the provision of services. And the new service economy requires emailing, text messages, and mobile calls. The smartphone is the new factory, and emailing and messaging are the new labor process. Marx foresaw solidity melting into air (Marx and Engels 1998 [1848]), but perhaps not in these dramatic ways. In iTime, one can work anytime/anywhere, freeing one from the cubicle but also allowing work to seep into every nook and cranny of personal life. Upside: freedom from the job site. Downside: no downtime. This accessibility is made possible by the new labor process and its corresponding commodity form: electronic output. One of the great struggles in the Internet era is over how to commodify the potentially free good of Internet traffic. Publishers understandably want readers to pay for what they are delivered onscreen.

For Marx, the commodity form enveloped labor, as it did all other commodities. Physical labor was bought and sold in the labor market, making the extraction of surplus value possible. Since the mid-nineteenth century, labor has evolved from manufacturing to office-based work, as the workforce has changed the colors of their collars, from blue and pink to white. Capitalist industrialization has traced a four-stage pattern. In the first phase, from about 1700 to 1900, the early factory system sprung from the growth of European and American cities. Technology was rudimentary and output minimal. The struggle to survive was intense. A second stage began with Frederick Taylor (1947) and Henry Ford's efforts to measure and manage industrial output. Ford initiated the era of mass production, using the assembly line to produce inexpensive goods in great volume, putting America on wheels and providing a model for other modes of production, eventually including agribusiness. A third stage began just after World War II and stretched to the late 1980s. This was the era of the "organization man," as capitalism entered a bureaucratic stage and the labor process began to shift from factory to office (Whyte 1956; Mills 2002). Educational credentialism helped stratify labor (Collins 1979). Finally, our fourth and current stage stems from the Internet in the late 1980s, which itself follows from the miniaturization of computing necessary for the United States to land a man on the moon in 1969 (Dyer-Witheford 1999).

This is a global era of laptop (or fast [Agger 1989a]) capitalism. Many people no longer need to dress for success in the white collars of the third stage (or their female equivalent). They can wear t-shirts, as they make their office at home, at Starbucks, even in their cars. This is the era of telecommuting (Leonhard 1995). Initially, the just-in-time production and service provision characteristic of this era tethered the worker to a desktop work station. But, with laptops and now smartphones, the connection between work and a singular physical space has been undone. The labor process has become mobile and workers now inhabit what I am calling "iTime." Labor in iTime has these features:

- Labor is largely writing, of emails, texts, memos, blogs.
- There are no clear boundaries between the process and product of work; professional work time is devoted to life on the screen, which bleeds into private communication as interlocutors transact business and chat, sometimes indistinguishably.
- Labor is not bound by either the business day or the office; it is conducted anytime and anywhere, thus effacing the very distinction between paid and unpaid work time.
- Labor is self-reproducing, as messages trigger other messages, seemingly endlessly.
- iTime oozes everywhere, driving out downtime.

iTime is consistent with, and hastens, the expansion and elasticity of the commodity form in late, laptop, fast, post-Fordist, postmodern capitalism. (These are all synonyms, suggesting simply different lineages of the concept of a global, Internetworked capitalism.) Marx used the term *commodity form* to capture the unique process of buying and selling in capitalism, including of labor itself. In *Capital*, he "unpacked" the commodity form to reveal relationships of subordination and superordination (Marx 1967 [1867]). The commodity is not pristine but emerges from struggles between capital and labor over surplus value as well as over control of the working process and working day. Marx implied that there is a region of life outside of commodification and the market, typically viewed, since the Greeks, as "private" life.

In the 1960s, feminists, New Leftists, gays and lesbians, and even hippies pointed out that so-called private life had become a contested terrain as people struggled over bodies and over the environment. This struggle has been extended to time itself, a terrain already identified by Marx as crucial for the development of capitalism. Marx's whole analysis of labor's commodity form pointed to a duration of unpaid labor time in which workers are, strictly speaking, not paid for their work. They would thus transfer enough "value" to the commodity that it could be sold for a profit. Marx argued that workers morally deserve that profit since it came off their backs, but Adam Smith and David Ricardo felt that it was obvious that entrepreneurs, who took the desperate risk of starting a business, deserved some return on their investment.

Marx understood that time, especially the boundary between paid and unpaid labor time, was crucial for understanding capitalism. He did not foresee the colonization of non-work time in later forms of civilization, in part because he did not foresee that capitalism could precipitate consumerism once people's basic material needs were satisfied. After World War II, so-called leisure time was actually spent briskly in the shopping malls and now online, as people, using credit, keep the engine of capitalism humming. Marx never expected capitalism to survive to World War II because he felt that its contradictions were so sharp that there would be a crisis of underconsumption (or overproduction), given the tendency of labor to be immiserated as capital was further centralized and concentrated. Marx felt that the "logic of capital" was inexorable, and he also felt that workers' class consciousness would be elevated once they were thrown out of work and read tracts such as *The Communist Manifesto.*

Marx failed to foresee two things: the state's growing ability to intervene in the economy in order to prop up the profit system, banking, and the poor, as was amply demonstrated during and after the Great Depression; and the culture industry's ability to resolve psychic crises, such as meaninglessness, anomie, and depression. Marx underestimated the resilience of capitalism, even if his analysis of the alienation of labor and of the crisis tendencies of capitalism was convincing.

I read iTime as the latest way of co-opting, coordinating, and commodifying human activity, enmeshing people in the microphysics of power, a grid that binds them to an everyday life lived thoughtlessly. Smartphones are engines of production, as I noted, and they are also commodities, requiring an initial purchase of a phone and expensive monthly payments. But labor's Internet commodity form also contains struggle, as Marx's original concept of labor's commodity form involved struggle between workers and capitalists over both control of the working day and of the disposition of surplus value. In a laptop capitalism, iTime or mobile time extends the commodity form of labor but also provokes resistance, as people battle over the control of their time, which is in effect control of the working and living process.

In my own university and many others, there is a move to put faculty on differential teaching loads, with "productive" faculty (in terms of research grant dollars and publications) earning lighter loads. Production quotas are being set—so many dollars or articles published per year. This is a perfect example of the expansion and elasticity of work time, now to be conducted anytime and anywhere using the prostheses of information and communication technologies. With budget crises, telephone landlines are being taken out of faculty offices, as faculty work offsite and outside the parameters of the standard Ford-era working day.

The problem in academia is that the life of the mind is not a business; intellectual serendipity cannot be structured in the way a sausage factory might be. Imagine the innovative research scientists at the Institut Pasteur in Paris in the early 1980s hunched over their laboratory benches worrying about meeting their academic production quotas. If they didn't have intellectual freedom, including free time, they might well not have discovered the HIV retrovirus, eventually tracing it to African chimps in the early 1900s.

This is not to deny that labor's commodity form under Fordism was grounded in time, as Marx showed that it was. Capital exploits labor through time, notably through the duration of the working day during which workers, in effect, work part of the time for free, "necessary" labor segueing into "surplus" labor, the ground of private profit. But in the Internet era, labor time stretches and bends elastically such that labor time extends beyond the eight-hour work day of the earlier twentieth century. Far from a labor-saving device, the Internet compels work in unprecedented ways and makes work's surveillance even more seamless.

Limitless Accessibility and Manic Connectivity

It is nontrivial that people are always available as they exist in iTime. This involves a certain deboundarying that could be said to be characteristic of the postmodern moment. Of particular significance is the deboundarying of the public/private distinction as mobile information and communication technologies allow for nearly limitless access and availability. The main modes of connection are email and text messages, and the work station of choice is the smartphone (for a discussion of cell phones, see Rippin 2005).

One cannot hide in iTime. Boss, colleagues, and family expect one to be available. Although phones that double as computers can be silenced, the mail and texts accumulate, driving the need to respond. The labor process increasingly involves simply responding to these interventions, sometimes accompanied by document attachments that are electronic versions of memos and other textual deliverables. Every post-Fordist proletarian knows that email is a burden as well as exciting. Writing and responding to it is work, but receiving it can electrify a dull routine.

We can read the urge to connect in two ways: People, especially kids, seek community through connection. They are isolated and alienated and want to remedy this. On the other hand, connection could be a neurotic (Adorno [1974] called it "damaged") impulse to flee one's self and one's situation. Immersion in the chatter of everyday life is a diversion from what is really going on—which is frequently nothing at all.

In any case, availability makes way for the compulsion to connect, not to miss anything. Smartphones are especially seductive because they bundle connection with entertainment, another kind of diversion. There is a simultaneity of texting and surfing that connects and diverts at once, enriching the daily lives of kids who are drowning in a sea of homework but who may receive little positive adult attention. To note that smartphones are babysitters is to suggest an obvious parallel to television, except that with phones kid are active as well as passive, reaching out, keyboarding, seeking diversion and information.

Adults sell smartphones to kids, and to themselves, in the way they shill for fast food. There are certain obvious efficiencies to be realized in a computer that doubles as a phone just as fast food can fill one up quickly and relatively inexpensively. The commodification of connection, entertainment, and information cannot be ignored; Apple is not a nonprofit, nor are the cellular services. Wants become needs as kids need not only smartphones but apps and monthly service contracts. Parents foot the bills lest their kids lag behind socially.

Time morphs into iTime as connection and diversion dominate one's waking hours. iTime is *mobile time*, time that is portable as well as elastic. To be ever-connected is the technological imperative of phones that double as computers and of laptops. Labor bleeds into life itself as one cannot risk being offline for long. iTime is time spent manically producing connection via email, texts, and blogs. The *New York Times* (Richtel 2010) ran a front-page story about a family that is always connected. He offices from home and they are always in front of the screen, many screens actually. He has an Internet-based business and his kids use the Internet for gaming and playing. He is rarely offline, and he repeatedly checks his email.

Mobile time is not only elastic in the sense that it extends into private time; it is densely compressed, weighing heavily on the person who always has too much to do, not enough time to do it. Mobile iTime feels infinite, never to be bracketed by beginnings and endings. I often dread my email for this reason, just as I'd be lost without it.

Self-help for manic connectors might involve limiting one's access to the Internet, perhaps checking email and texts only a few times a day. But what I am calling manic connectivity is imposed by the expectations of others, who resent being ignored. Where labor involves connection, to be disconnected is a kind of dereliction. And, in any case, connection is addictive—precisely the way in which social structures implant themselves via the consciousness, sensibility, and apparent will of otherwise rational people. In short, connectivity is structural; it transpires almost above the heads of the busy iTimers who are positioned by labor's Internet commodity form.

When I grew up, in the late 1950s and through the 1960s, time was relatively immobile. We could connect using intractable landlines, and we watched a rudimentary form of pre-cable television. Visitors came and went, and we played and socialized in the neighborhood. When we left home for school or play, we connected only face-to-face. And there was mail, which we devoured for the surprises it could bring! This isn't simply an ode to an unsullied era, when time didn't travel with us but stayed behind. I couldn't use the Internet to do research, nor learn the world quickly. But I wasn't compelled in the way people are today, and I still had my share of community. I played, hung out, dated, just as kids do today. My nighttimes were less fluid and probably ended earlier. Even television shut down, with the infamous test pattern imprinted on our TV screens. I learned early on that little that was good happened in the middle of the night.

The mobility of time, afforded by smartphones and other such technological prostheses, renders history immobile, even irrelevant. Time's mobility spells immersion—in the quotidian, routine, ever-the-same. It is nearly impossible to consider world history when one is so involved in one's world that there is no distance from which to contemplate past, present, and future, to think theoretically, in other terms. Although one can surf the web using the phone, the Internet is famously shallow, with Wikipedia its epistemological emblem. And few smartphoners use their phones for deep or even shallow research into theory but rather simply for entertainment, which occupies them during the downtime involved in synchronous text messaging.

This frenzied literary activity of electronic connectivity, situated in mobile iTime, need not be banal; that it is reflects the dominance of the culture industry, which diverts us from war and peace to the latest Hollywood gossip. Twitter in its very name connotes trivia and ephemera. The Internet is a massive culture-industry machine, saturating people with the flotsam and jetsam of the damaged life. We are thus diverted from our own alienated everyday lives and from the larger questions of history and utopia. Immersion and diversion occur together; indeed, the greater our investment in ephemera, the more diverted we are from what is really going on.

For iEpistemologies, there is no "really going on." There is only perspective. The mobility of iTime is matched by the mobility of truth. Postmodernity is iModernity, simulated, stretched, immersing, diverting, compulsive. However, I'm not an iModernist but a modernist, even if not of the kind imagined by Descartes, Kant, even Marx. However, Marx anticipated the melting of solidity into thin air as capital colonized not only the world but consciousness, an insight that made possible the Frankfurt School's concept of domination and Lukacs's reification. My own idea of an oozing textuality that seeps out of its covers into the world itself, compelling the *amor fati* (love of fate) as people experience the world as necessary and inevitable, stems from Marx's pre-postmodern anticipation of mobile time and a chimerical reality. Marx did not pursue these ideas because he was convinced that the rate of profit would fall and massive unemployment would lead to the revolution. It hasn't, in part because the melting described by Marx continues to dissolve history and utopia—exactly what Marx and Engels wanted to combat in the *Manifesto*.

I suspect that Marx would have theorized iCapitalism as a horizon of unprecedented conformity, control, and compulsion and yet fluid, open to new interventions and organizations. His and Engels's own *Manifesto* could have been released electronically, posted, blogged, emailed. Whether it would have disappeared into the thin air of iSpace or galvanized people to seize first the literary day and then the political day is again unknowable. One cringes at a wiki version of that revolutionary call to arms. But then who is to predict reliably the fate of missives launched into cyberspace—an electronic public sphere or simply an extension of the culture industry? It is clear that social media can be transforming, as the Arab Spring and Occupy Wall Street demonstrate. Hacktivism and cybercommunity organizing occur nearly at the speed of light, suggesting that globalization (via the Internet) contains, dialectically, the possibility of its own negation.

Telling iTime, Texting, and Tweeting toward Utopia

Time is above all linked to work, as Marx profoundly noted as he analyzed the production of profit during the working day, which he divided into necessary and surplus labor time. iTime vastly expands work, against the predictions of postindustrial society theorists such as Bell (1973). I am suggesting that smartphones and laptops are the work stations of post-Fordist capitalism. They tether us to work; they become the sites of labor. They make us available. And they lead to mania, as we constantly check our connections. The upside of being anywhere/anytime is the multiplicity of opportunities opened up by globality—cultures, interlocutors, information. The downside is the transformation of history (Hegel might have capitalized it as History) into iTime—an eternal present that has no depth, nor the dialectical possibility of utopian futures. People who exist in iTime are almost never utopians or radical critics. They accept the world for what it is—the people on their contact list, their bookmarked URLs, the consumer possibilities that exist online. Only by stepping out of iTime's compulsive connectivity can we resist flattening History—a present grounded in the past thawed into its many political possibilities—into what "is" or appears to be.

Postmodernity is characterized by History that dissolves into everyday time. Modern narratives of progress, whether Marxist or neoliberal, preserve the possibility of History as the emergence of major change from the temporal experience of people's everyday lives. These are the electric moments when history opens from a seemingly blocked present. For me, many of these moments occurred during the 1960s, when we, especially the young, knew that these were heady times never to be forgotten. We marched against the war and racism, and for peace and equality. We sang and danced these times, uniting mind and body. I recently published a book about these times, when lived history and History blended, which was spoken through the voices of the leaders and participants of black and white movements, women and men (see Agger 2009a). Inevitably, the story was my story, a memoir really, as I recalled my own teenage years and my parents' involvement in the New Left struggle. Memoirs are tricky because one must disentangle one's own time from Time, the merger of history and History.

I often wonder whether the iPhone generation will ever have these experiences of History opening up, as it did for me. I end my book with memories of walking a desolate beach on the Oregon coast as President Lyndon Johnson withdrew from the presidential race, leaving the field open for Bobby Kennedy, brother of the slain President Kennedy. I reflected on my own future as the future of our country and world hung in the balance of fast-breaking current events. A couple of months later, a few days after I stood near Bobby as he delivered a speech in my home town, he was assassinated in Los Angeles after winning the Democratic primary in California, dooming the peace efforts of the Democratic Party and ensuring that Nixon would triumph, thus beginning a very long reign of the hard Right, which also covers the Thatcher years. My walk on the beach quickly led me north to Canada, where I avoided the draft and the political turmoil engulfing the United States.

We boomers are awash in nostalgia. It is important to put ourselves in the shoes of the young, not assuming that they are apathetic politically and can never live History in the ways we did over forty years ago. It is easy to dismiss what we do not well understand, especially our multitasking, slouching, slacking, overstressed kids. We need to try to crack their code without assuming that their text messages can never measure up to our manifestos.

Laptop capitalism contains an interesting dialectic as people possess the means of writing, criticizing, organizing. Obama was elected using Facebook and other electronic means of organizing and raising money. Although the flattened, de-historicized iTime of a manically connected eternal present immerses people so deep in everyday life that they don't question their strange participation in the electronic grids of power and capital, the new means of production are also vehicles of critique and consciousness raising.

Children are examples of the dialectical nature of texting, chatting, emailing. They hide their phones while they spend the boring school day texting, using slang to resist adult power. Schools are prisons, and phones allow the inmates to stay connected by tapping on the walls of their cells. Writing teachers lament the decline of discourse, but kids write furiously, even if their homework is uninspired. The Internet allows kids to connect, criticize, energize, agonize— all of which they do in writing. It may be a short step from using Twitter and blogging to writing the galvanizing manifestos of yesteryear that helped move the world forward, such as *Rights of Man* (Paine 1985 [1791]). A manifesto is a blog with a plan.

Most kids share the postmodern aversion to politics as the boring grand narrative of their parents. And yet they networked for Obama, who embodies postmodern cool and doesn't condescend to kids. An iPolitics is possible that turns iTime into a more considered and critical moment of a History about which we can tell big stories of possibility and passion. I have argued that the postmodern needs to be postponed and instead that we need to fulfill what Habermas (1987a) calls the project of modernity instead of abandoning it. We can use technologies of communication and connection to create community, just as we can abundantly be used by them—spending endless hours chatting about nothing, what Adorno condemned as chitchat. I am composing the paper on a laptop, and once my netbook arrives, I'll transfer the file to it, probably by using my thumb drive. I texted our son's tennis score to my wife before I sat down to write these sentences. And I edit an electronic journal called *Fast Capitalism* (www.fastcapitalism.com) that considers the impact of rapid information, communication, and entertainment technologies on self, society, and culture in the twenty-first century.

That labor in iTime is largely talk suggests the possibility of discursive insurgence, of critique. Although companies try to read their employees' electronic messages, this is not a seamless process. iTime is messy and difficult to control. Kids have elaborate networks of resistance. Adults can resist in many of the same ways as their kids. Foucault's panopticon, a perfect prison in which surveillance is obviated by self-surveillance, is somewhat pre-electronic. Although Internet cookies can track users' preferences and surfing, there is a level of protean writing difficult to constrain by Big Brother. The volume is too vast and time is scarce, even though iTime bends and expands beyond pre-postmodern labor time, which was rarely more than forty hours per week.

Some of the most interesting iTime is late at night and early in the morning. Kids haunt the night, after school, dinner, and homework, suggesting a "vampirization" (Gloria Goodale, personal correspondence) necessary to survive in the daylight world. Sleep deprivation is chronic, among both kids and their parents, because the flickering screen beckons and because days have become so long that there needs to be downtime. Only stoics spend this downtime offline or away from television. The image is seductive and the need to write elemental. Perhaps this is only to notice that people seek connection, and phones and computers reduce the friction involved in forging connections nearly to nothing. Whether these are "real" connections, of the traditional face-to-face kind, is perhaps a question not well understood by denizens of postmodern time, for whom most meaningful interaction is electronically mediated.

Kids' time stretches and bends; they are genuinely postmodern. Our time is linear; we are modern. This is more than the clash of generations. It represents new ways of doing everyday life, community, work, education, leisure. The compartments dissolve, as do boundaries, as kids carry their iPhones with them, suggesting a prosthetic metaphor. Not to be "on" is to be asleep. Work is everywhere, as is interaction. There is no disconnection.

For the connected generations, utopia—the good society—is 4G connectivity, while for us, their parents, utopia is a reprieve from connection and accountability. We don't want to have to check our email or texts, even as we have become habituated to doing so. I remember an early Internet era message at a former university, advising us to log on to check our email at least twice a day. I was instantly ambivalent: I could see the utility of electronic connection, but I resented having to be plugged in. Today, many check twice a minute; they are always in receive-and-send mode. They miss nothing, but their time does not lead forward into a qualitatively better future.

Slowing Down iTime

Lyotard rejects grand narratives of unfolding time because they seem to him potentially authoritarian. I characterize utopia as "slowmodernity" (2004a), blending a pre-modern return to nature with high technology. I share a distaste for large stories that sacrifice minds and bodies to the cause. I'm New Left in that I want change to pass through, and transform, everyday life, including the very concept of everydayness. Hippies during the 1960s wanted everyday life to harbor transcendence, not to be set apart from it. They, too, valued connection, but of the face-to-face, body-to-body kind. In a way, they were pre-postmodern, as was Situationism, from which Marcuse (1969) drew in *Essay on Liberation*.

Marcuse wanted people to embody, and become, "new sensibilities," working, playing, loving, living differently. They would seek the "promesse de bonheur," the promise of happiness found within the body itself—a "rationality of gratification," as he termed it in *Eros and Civilization* (1955). This rationality would give rise to new sciences and technologies through which people would play with ideas and techniques. Was he describing the Internet and its organon of the smartphone? 4G connectivity allows for speed, flexibility, reach. One can connect, research, write, nearly at the speed of light. The idea that everyone is a writer is potentially revolutionary. Derrida (1978) noticed that every reading in effect "writes" a different, perspectival Shakespeare or Marx. Word processing is the means of literary production allowing reading to go public, eroding the hegemony of official knowledge production centered in universities and in corporate publishing.

These are positive ways to view the Internet and the busy literariness of those who spend much of their time in front of the screen, whether anchored to the desktop or mobile. Derrida's insight into our inveterate writerliness suggests a democracy of readers and writers, none of whom can claim special privilege. No matter how much people try to commodify the Internet, web pages and blogs slip through the nets of control and capital, staking out subject positions from which people can do valuable literary work and even organizing. Obama might not be president were it not for Facebook and cell phones. MoveOn.org represents the grassroots political implications of an electronic public sphere, suggesting that iTime pushes against the Internet commodity form and has a democratic intentionality that cannot be easily contained.

Perhaps this is simply to notice that there are many Internets, diverse forms of electronic connection, and eclectic versions of iTime. Obama has a BlackBerry, as does my physician, as did my son. Their time is flexible, Mobius-like, but in different ways. They are all in their way postmodern, but in the case of Obama and my physician they still have one leg firmly in modernity. My son will grow up never having known boundaries between public/private, uptime/downtime, labor/leisure. He will only have done his homework with music blaring and perhaps various web pages open at the same time. His friend texts him while he does his math, and she sometimes helps him solve his problems; he is not driven to distraction but cannot focus without his electronic accoutrements.

Postmodernists banish utopia as a hubristic, even dangerous, construct. But the boundary between the modern and postmodern is blurry. I have argued in *Postponing the Postmodern* that we are not there yet, mired as we are in a late, fast, or perhaps even postmodern capitalism (Agger 2002). As such, I agree with Habermas that we should hold onto the idea of modernity and fulfill its project, instead of abandoning it. We thus embrace reason, peace, freedom from hunger, tolerance. But globalization, the Internet, and the dissolution of the Soviet Union call into question earlier metanarratives of modernization. I hold onto those large stories because my Marx is he of the early manuscripts and, in mediated form, the New Left, blending with the Frankfurt School. Yet one can be modernist at a moment when all that is solid melts into air and clocks ooze a different, nonmillenarian time. This is iTime but it can be clocked, must be clocked, in the modernist way, with downtime as well as uptime, night and morning, boundaries.

This is the heart of the matter: Postmodern iTime is not boundaried in the way that modernist clock time is, not to mention pre-modern natural time (based on sunrise and sunset). During iTime, people can communicate asynchronously, through saved messages. And they are available at all hours of day and night, incorporating their smartphones prosthetically. The body-subject, as Merleau-Ponty (1962) called it, is transformed into a phonic body-subject, as orality is broadened and transformed into speaking/listening/typing/texting. Human communication is not simply mediated electronically but incorporated into the self in a way that turns us into cyborgs, about which Haraway (1991) and others have written extensively. The human body-subject takes on an extra dimension as we live life through the smartphone, which becomes a mode of subjectivity itself. It is a mode of subjectivity that includes intersubjectivity, a central insight of Husserl's phenomenology (Husserl 1970).

iLife is inherently intersubjective as boundaries between self and others and self and world dissolve. Self and others intermingle through the form of life of the text message. Apps are ways of allowing the world in, connecting us nearly instantly to information and entertainment. This is not a technological-determinist argument because there are different ways of using, and being used by, smartphones, computers, cells. But I am identifying boundarylessness as a central tendency of iLife lived through iTime.

Are boundaries good in themselves? There were too many boundaries before modernity, between villages, castes, regions, nations, religions. The project of modernity could be described as a deboundarying in the sense of destratifying and globalizing. The postmodern takes this further, too far in my view, effacing the very boundary between life and iLife, time and iTime, and thus reducing the self to a "subject position," the apotheosis of modernist reason.

My kids and students are positioned by their smartphones, scarcely understanding that they could use, and be used by, them differently. Utopia is a new app, jailbreaking their phone, texting their girlfriend or boyfriend. We elders can help them think about utopia in ways that trade on early Marx, the New Left, feminism. This will emphasize the possibility of a new everyday life achieved through everyday life itself, which links up with larger social movements as what Hegel called particular and general are reflected in each other. We won't postpone liberation to a distant future time, nor sacrifice minds and bodies along the way. Our lives will be "prefigurative," modeling the future in the present as we develop nucleic forms of life, including discourse, that contain the future embryonically.

Kids don't talk and think this way because they are ensconced in the eternal present of the postmodern. They are especially oblivious to the past, even though they learn historical factoids in school. They don't think deeply about the connections between past and present, preferring to live iLives through a phonic body-subjectivity. iTime doesn't include the past, the muck from which microcomputing and ICTs developed slowly. And without a past, there can be no future discontinuous with the present—and hence no utopian imagining. We need to teach or re-teach our kids the 1960s as a reminder of how college students and other young people changed history by changing themselves into sentient, peaceful, and democratic beings—precisely the agendas of the early Student Nonviolent Coordinating Committee (SNCC) and early SDS.

When I screen videos about the 1960s, for example about Kent State or the Weathermen, my students are astonished that this was such recent history, involving their own boomer parents and teachers. Kids then were immersed in narratives of historical unfolding. They were political, not phonic, body-subjects. Connection took a backseat to community, consensus, and the causes of the day. Time marched on, as people marched toward a different, better future. The counterculture and hippies, the apolitical moment of the sixties, anticipated iTime in the be-in, a public non-event celebrating the present as a plenitude of being. Music and dope positioned them in the eternal present of the be-in and concert, much as smartphones do today. Acid was akin to what today we call an app—helping one transcend from here and now to a transformed mode of experience, if not of a transformed world. The more political kids of SNCC, the Congress of Racial Equality (CORE), and SDS were transforming the world, even if they paused to smell the incense.

iTime stands still just as it is mobile; that is the problem politically. What Lukacs termed "reification" freezes the world into an eternal present. History is obscured, and so is the prospect of a different future—perhaps a utopia. Jacoby (1999) comments that the decline of utopian thought and writing is a huge casualty of the timeless, frozen present. I have written of the decline of discourse, and now I understand the role of time, especially, as time is produced by those who profit from frozen time and who dehistoricize time in order to reproduce it. The Frankfurt School well understood that social amnesia (Jacoby 1975), a political forgetting, closes off utopia by freezing contemporary experience into a plenitude of being. Their critical theory could be read as thawing frozen time, reading the present as a molten, dialectical moment at once containing the past and auguring something new. Hegelians call this the dialectic, and iTime is the suppression of dialectical forward movement.

The risk in all this is to re-install a master narrative of linear time, of ineluctable progress, that sacrifices lives to a supposedly universal historical destiny. But iTime is not an adequate response to Stalin or the hard Right. Forgetting the past dooms us to relive it, as the great bearded one remarked! Some kids may archive their text messages and blogs, but who reads backward to last month, let alone last decade?

The challenge is to make the past come alive as the possibility of a different, better present and future. We counter the forgettings and freezings of iTime not by reviving the past but by opening up the possibility of a better future, and a future dialectically connected to the present in the gestures and agency of people's everyday lives.

That is the way I teach the sixties—as an examplar of present and future radicalism. To get there, one needs to thaw the past, especially using a story-telling mode. That is the great contribution of a postmodern ethnography that retells the past, especially in its own voices and using images amply. This will especially tempt the young who write their own lives, and who are posttextual, dwelling in imagery and icons. We need to rethink not only time but writing, history, sociology as ways of demonstrating the possibility of utopia. Writing time is one way to slow it down, from which we can project better time. The challenge is not to get the past straight for there will always be confusion, ambiguity, and contradictory stories told. Rather, the challenge is to demonstrate that the present's inherence in the past (its "historicity") is an opportunity to make new history. The sixties are so compelling as a story-able moment because the New Left, white, black and brown, male and female, insisted that social change must pass through the self and everyday life. Only as the decade petered out did the Weathermen and the Black Panthers, for understandable reasons given the Right's counterrevolution against the earlier peaceful sixties, tend to obscure the self and participatory democracy in favor of armed combat—never the way to go, especially given the amassed power of the state.

There is discursive parallel between then and now. Kids today use the same word, "cool," that we used then. Cool is someone who can be trusted; it signifies solidarity. I contend that iTimers yearn for time (even Time) even if they cannot name or identify the political protagonists—not their fault because conflict has been cloaked in the meantime. Bull Connor's police dogs have been curbed, and a hot politics of revolution and counterrevolution has been muted, partly because of the collapse of the Soviet Union (and hence, Americans conclude, of Marxism) and partly because the Right is, if anything, even more hegemonic than during Nixon and Reagan. Neoliberalism reigns supreme and globalization is the new framework for American empire.

Perhaps it is enough to note that iTime is contested terrain as time remains a crucial factor in the struggle for freedom from coercion and control. Our kids may not talk this way, and so it is for us to provide them, but gently, a vocabulary of resistance that contains an image of utopia—for me "slowmodernity." Whether this slow/fast image will contain the smartphone is somewhat beside the point; people will be smart again, and their phones/computers will be mere instruments, not a deep part of their being. This is a re-Kantianization of sorts, even though

I have criticized Habermas's (1971) *Knowledge and Human Interests* for relying too heavily on Kant. Although I want to believe that Marcuse's vision of new science and new technology blurs the boundary between nature and humanity, the new info/comm technologies are so compelling that the Luddite in me surfaces and I hold onto Kant and Habermas as bulwarks against the sheer usurpation of reason. Or perhaps one could imagine "playing" with the iPhone in a way that expresses the life instincts, while avoiding a Foucaultian disciplining. The optimism I felt when I read *Essay on Liberation* has been somewhat constrained by the subsequent half century of political regression and technological colonization.

In the next four chapters, I explore youthful alienation, rebellion, and utopias as expressed in kids' attitudes toward authority, adults, schools, time, and schedules. In Chapter 7, I address slacking.

CHAPTER 7
SLACKING TOWARD SLOWMODERNITY

Slackers is a new term for an old phenomenon. Like beats and hippies before them, slackers are young people who aren't career-minded, instrumental, productivist; they do not seek to emulate their parents or other adult role models. They may be into music. They are expressive, and some use drugs and alcohol. They tend to congregate in countercommunities and to speak in the shorthand code of texts and posts. Slacking is defiance, a rejection of the adult performance principle.

The term "slacker" was popularized in a 1991 independent film by Richard Linklater. He did not intend the term to have negative connotations, again tapping into the long industrial-age tradition of rebellion and resistance about untoward productivity expectations. The film, shot in Austin, Texas, depicted a day in the life of twenty-somethings in that area.

Slackers are reacting against performance expectations, time pressure, compulsive evaluation, homework, career and life planning, delayed gratification. This is not necessarily to read slacking as utopian, the grand gesture glorified during the 1960s. Slackers, unlike hippies, do not always intend for their private behavior to become a universal rule even though their behavior may bear that implication. Hippies wanted to eliminate toil, war, hierarchy, structure. The sixties counterculture was related, if not identical, to sixties social and political movements. Many grew their hair long and smoked dope just as they burned their draft cards and marched against the war and for civil rights. Marcuse (1972) lamented the turn away from politics by sixties movement people who decided that burning their draft cards was futile and chose to burn incense instead and decamp to communes.

Progressive adults from the sixties try to be cool parents and grandparents, appreciating their kids' rap music and their social networking. But few of us, if we are honest, embrace the postboomer generation's inattention to politics and social movements. Even for those of us who grew our hair long and smoked dope, it is difficult to understand slacking as the pre-political attitude it may be. In rejecting the utilitarian, slackers reject all utility, rendering them vulnerable in a dog-eat-dog world with a sagging economy. Although we may bemoan the performativity of more conservative parents, driving their kids to distraction with a stress on accomplishment, slackers command little sympathy simply because we are purposive and they are not. We want to get the job done, even if that isn't earning our MBAs but rather electing progressives. We are driven, and we drive our kids crazy with our war stories from the sixties.

I array the pre-modern, modern, postmodern, and now the slowmodern, as I term it. The slowmodern combines elements of all, putting the brakes on accelerated life in retrieving agrarian rhythms but also retaining, if transforming, literacy, democracy, abundance—modernity. Fast capitalism can be slowed without turning back the clock, which is never possible anyway. Opting off the fast track does not mean being off track; there are other tracks, other ways and qualities of life. Petrini's (2003) "slow food" movement, involving eating slowly in convivial settings outside of the corporate mainstream, is a good example of the slowmodern. You can be a hippie and have a roof over your head. You can even join a social movement.

Beats, hippies, and slackers have in common an aversion to structured time and rigid lives. They reject bureaucracy and the bureaucratic work setting. They do not like rigidity, constraints, subordination. They are libertarian. One might view all of this as a rejection of "productivism"—the assumption that the person's worth is tied to his or her output (of whatever—sales, memos, homework, base hits). Luddism from the mid-nineteenth century, during the Industrial Revolution, anticipated these twentieth- and twenty-first-century subcultures. Industrialization was seen as inhuman and a violation of nature. Eventually, time and motion in the factory were linked by Frederick Taylor (see Bendix 1958), who authored the approach called "scientific management." People were cogs in machines, or just machines. Bound up with Luddism was a high priority placed on nature and naturalness—hence, the organic food and cooperative movements of counterculture people during the sixties.

This is a history of rebellion and even radicalism. Slacking does not call up images of the rejection of technology, workaholism, and the domination of nature. However, it can be read as a youthful expression of resistance and resentment situated firmly in this more venerable political movement. Slackers may not understand this or intend it, anymore than incense-burning, pot-smoking hippies at sixties be-ins recognized that they were engaging in a Great Refusal of bourgeois values and capitalism. High-school kids, in rejecting dominant value systems and especially their own regimentation, are pre-political; they could be mobilized and come to recognize their own values and sensibility as political, indeed as transformational. Marcuse in *Essay on Liberation* termed this nature-loving, nonproductivist mind-set the "new sensibility," a sensibility that rejects capitalism and favors a more erotic society in which human needs do not despoil nature and lead to authentic human gratification.

Slacking has a place within the history of protest against capitalism and its workplace, lifestyle, and even corporeal expectations. Battles are waged today about high-school dress codes. Many parents and administrators propose a tighter dress code or even school uniforms for these nearly-adult teenagers. Our local school board voted down standardized dress—permutations of uniforms, in various pastel colors—by a close vote of 4–3. The kids were relieved because they want to express themselves, learning who they are even if they commit fashion mistakes and offend adult sensibilities. This is racially inflected; our public school system has many minority kids, some of whom (boys mainly) dress in urban styles including baggy pants that sag below their butts. This dress is protest—against The Man, authority, Puritanism. It is a pre-political statement, and authoritarian adults recognize it as such.

There is already a district dress code in place. Sagging pants and the exposure of too much skin on girls could be sanctioned within the current code. But adults are piling on, trying to make a point: Slackers, exhibitionists, gangstas will not be tolerated. One surmises that the 2008 Obama election might have heightened racial sensitivity, which, in a mediated way, is expressed in the desire to tighten up dress standards in schools. Or perhaps adults—administrators, teachers, parents—sense that kids are amassing, a slumbering pre-labor force on the verge of rebellion. The slacker's attire is the military uniform of this people's army.

During my high school years, culminating in 1969, I wore nickel pants purchased at Goodwill. Those were the days. Even politicos like me dressed down, like hippies. The New Left and counterculture blurred; we all loved the music, incense, even dope—a telling word. A segment of us went further and marched against the war and for civil rights. The elders hated our hair and Goodwill attire. The generational wars continue, but I am tempted not to view this as a fact of life. Connecting then and now is war itself—first Vietnam and then Iraq and Afghanistan. In Vietnam, over fifty thousand American lives were lost; in the Middle Eastern wars, over six thousand have perished. And there are culture wars now, as there were then. Generational conflict was political conflict, over values, society, the future. We had a uniform, too, although it was less uniform than the slouching jeans full of holes, which now are on sale already distressed at Abercrombie & Fitch.

It is easy to view slacking as postmodern, a symptom of generation X, as Douglas Coupland (1991) named it. We from the sixties were thoroughly modern; we believed in absolutes such as Truth and Justice. And we were sure that we were on the side of the angels. Today, all sides are wrong, all perspectives flawed. Postmodernists even term the individual self a "subject position," adding to dehumanization. But I prefer to locate slacking in a venerable tradition of resistance to modernity that is more countermodern than postmodern. "Slowmodernity" blends and blurs the positive features of modernity—universal literacy, high technology, global communication—with the positive features of the nonmodern and pre-modern: small-scale technology, direct democracy, organic farming, the redemption of nature. The postmodern gives up on modernists' sweeping dream of a universal Reason, which extends from the Greeks to Hegel and Marx. I am not a postmodernist because I believe that the project of modernity has yet to be completed and because I am a humanist. Viewing the self as a subject position is alienating.

Now most slouching kids, skipping school and skimping on homework and college applications, have not read Ginsberg, Kerouac, Kesey, or even Coupland. They do all the discourses I have been discussing, from Facebook to texting to blogging; their time spent with books, especially existentialist anthems, is scanty. Slacking is not ideology or theory, although it can be theorized. It is a posture, a gesture, an embodied stance. Where Shapiro, the running guy, runs across the country, apparently going somewhere in modernist fashion, slackers slouch in place, barely getting off the couch. Shapiro learns that modernism is chimerical because the map promising a destination is in fact in our head; there is no end, or perhaps the end is everywhere, in Utah and Ohio as well as his hometown of New York City.

Shapiro trained for his "transcon" by running marathons and ultramarathons; he was already a long-distance hauler. Slackers eschew training as a capitulation to adult utilitarianism: Sacrifice now for rewards later. Beats, hippies, and slack-ers reject not only the rewards but the self-abnegation required to achieve them. It is easy to view this as laziness, and for many it is. But we are all high-energy units; lassitude is learned, it is an outcome of too much adrenaline and cortisol, a byproduct of stress. It may well be that slouching and sliding by are effects, not causes—kids having been beaten down and whittled away by adult expectations. Perhaps slouching is the posture of alienated kids, who, when they theorize it a bit, decide to become card-carrying slackers. They evolve from silent resentment to uniformed members of the pre-labor force's people's army. Their enemy is adult ambition, which, like the beats and hippies before them, they view as a bourgeois trap.

The beats had poetry and the hippies music. Slackers have music, but they thrive on the Internet, where they build community, emote, express, agonize, and game. Adults, even when online, occupy different corners of cyberspace. They also trek the information superhighway, but they are not cybercommunards. They seek love, sex, good deals on Craigslist. Their kids upstairs are burning up the cable paths, also in search of love and sex; but above all they inhabit a kids' world in which adult values and discourse do not hold sway. All kids slack when online; they shun homework, refuse to practice their instrument, don't take out the garbage. They are writing themselves, easing the wounds of the day. Few adults get that they are causing the kids' suffering.

Slackers and the Utopian Impulse

What do slackers want? They want less—expectations, scheduling, workload, social pressure, haunting by their futures. They want to blend the modern and pre-modern in their own lives, not in a postmodern way, where there is no stable ground, but in a "slowmodern" way, blending leisure and work, play and produc-tion. This is what the hippies sought during the sixties. They wanted to end the war and to dance around the maypole. The Diggers may have called up an earlier Luddism, but most hippies realized that when Bob Dylan electrified his guitar this meant progress, and psychedelic light shows backing the San Francisco bands were products of technology. That they wanted love and peace did not mean that they were regressive. They merged progress and regress, suggesting a different telos of history, a different narrative. We would no longer move from agrarian to industrial to postindustrial and postmodern but, beyond the modern, we would blend the best of each phase. This would not end modernity—technology, industry, communication, medicine, globality—but it would fulfill it, rescuing humans from the maw of a machine age.

Few slackers talk this way. They resist theorizing, especially their own lives. The thin discourse of the chat rooms and message boards does not drink deeply of Plato or postmodernism. But those are just words. They can be learned. That is why, if slackers manage to get to college or a good bookstore, they learn a whole new vocabulary, a whole new way of seeing themselves in relation to the world. They shed the uniform of the people's army, reinvent themselves, get a major in the arts or sciences and become the people we became after we decamped the sixties. Some of us became professors, others became professionals, teachers, artists, or stayed home to raise kids. We never left the sixties; they are in us, in our mistrust of authority, our rebelliousness, our creativity, yes, our hubris.

Here is a difference. Slackers lack hubris; many seem beaten down, like prisoners. The beats and hippies were full of themselves. Slackers often seem overwhelmed, especially in the high-school hallways. They have been incarcerated in what they experience as prison for over a decade, and it is getting old. And their postures and gestures don't help; adult authority is bound to be challenged by sagging pants and what adults, especially in authoritarian locales, read as disrespect. Slackers could use political theory in order to energize themselves: Pirsig, Camus, Fanon, Freire. Perhaps early Marx and Marcuse, as I was reading in those rebellious years. Such reading would help them situate themselves as strangers and rebels, not simply misfits. This is not the required reading kids are doing. Yet, as I noted earlier, kids read and write outside the lines, perhaps cultivating themselves and thus initiating social change in ways unrecognizable to many adults, even ex–New Leftists from the baby-boom generation. Wherever we find frenetic literary activity, we can detect, if not fully understand, the heartbeat of dissent and utopian reconfiguring. This is perhaps a generational trope, as parents fail to remember that they were young once.

Derrida helps us understand that writing is literary work, but reading writes, interpreting and in a sense creating the text. The personal narration of one's day and night is compelling to the storyteller and perhaps to loved ones. Some slackers read. In my day, the slacker book of faith was *Catcher in the Rye*, followed by *One Flew over the Cuckoo's Nest*. But these were not text messages or blogs. They were studied modernist works, not dashed off but carefully and slowly wrought. Their artifice was concealed in their fluid style. Kesey did not resort to emoticons. He labored over his prose to make it seem effortless, which is not exactly the same thing as noting that he had something to say. The slouching literati have plenty to say, but most of the words are lost in space. They need anchors for their writing, which they will only find in reading.

Slackers and the New Writing

My college library, not atypical in this, is crowded with people word processing, messaging, gaming, surfing. Few kids are reading, whether Pirsig or Plato or their chemistry textbook. But reading is going on, if a kind with which we elders are unfamiliar. They stare at their screens, engaging in the synchronous back and forth of messaging and chatting. They are talking, but of a type that seems like writing. It is as much like writing as to appear to be the kind of work Salinger was doing, but it is as little like writing as smearing is like painting. It is not inferior, just different. I often wonder what the kids have to say to each other.

My own kids and their friends open a window on this. They talk music, movies, love, parents. They seldom talk about the world, which weighs heavily on them. Their talk is connection and escape. It relieves them after a long school day and parents breathing down their neck. It is not productive in the modernist sense of generating output, such as a term paper, story, or novel. It is process, although, like more anchored literary work, it can be archived for further review. Adult email is in between the modern and postmodern. People, in a litigious mood, keep past email chains in order to hold people accountable or get themselves off the hook. They may even print out these messages and file them, combining postmodern pixelation and modern memorialization, pixels transforming into pulp. For most kids, texts and even blogs are not meant to be savored and are soon forgotten.

But they imprint themselves on sender and receiver. Hours spent writing—texting, messaging, posting—leave their mark. Kids are different for having opened themselves to others. Facebook is a metaphor of a children's public sphere. It is their town square, their message board. They are boundarying their world, keeping adults out. This is a defensive maneuver as adults encroach and quicken the pace of youngsters' lives. Perhaps slacking is best defined as wanting to prolong childhood; in spite of James Dean's preternatural cool, he defied adult authority and expectations. Don't tell this to the students slouching on the school ground and sprawling at the back of the class. They posture as wiser than their years, even as they want to regress to a time before adults and authority.

This worldliness is protest: Kids want to keep the world out, or to slow it down. They don't want to build a resume and take the practice SAT while still in junior high. All that can wait.

Adults understandably want slackers to shape up. We remember hippies, or perhaps were dancing around the maypole ourselves at be-ins. Slack is the opposite of tight, taut—taught. We want our children to have good and secure lives. A college degree is necessary to join the conversation—to obtain a job and a mortgage. It is the entry-level credential, which is a shame because many good jobs don't require really college. One reason poor kids turn to crime as a career and drop out of high school is because there are few well-paying jobs for the non–college educated, as there were just after World War II and earlier. Poor kids feel that they have few options to crime, imitating their cooler and more risk-taking peers.

Slacking is a leaning, a gesture, an ensemble of resentments and resistances. Alienation can be read for its underlying cause: the lack of meaning and of liberty. Kids who skip school, drink at lunch, and flaunt their defiance—all the while messaging their time away—are not in themselves a transforming class, possessed of political agency. But we can read their alienation, empathize and redirect them—to unstructured liberal arts colleges, into writing workshops and bands, toward literary and creative careers. This is a lost generation only if we abandon them. For many of us from the sixties, these kids are the way we were.

Teachers, and I include myself, are tempted to teach to the kids in the front row who show up for class. But in the absence and silence of kids in the back, the slouchers and daydreamers, lies human potential. Sometimes the keenest ones are not among the sycophants who sit up front and grub for grades. It takes effort to listen, to reach out. Many kids lack compelling adult models. They may live in broken homes, or, if they have both parents living with them, their parents may work so hard that they have little time or affect for their children. Teachers are overburdened and their classes are too large. There are few safety nets for the kids who aren't sitting up front, whose names are easy to memorize and who visit after class. The marginal ones remain marginal unless adults intervene.

And so the beats, hippies, and slackers cannot be considered in isolation from the adult world that spawned them, against which they are rebelling. If teachers can identify these kids early, they can make contact and stay in touch. Scratch a slacker and you find a kid who is wandering and searching. Often they are alone, isolated from adults. It is easy for schools to be punitive, but more important for them to be compassionate and to afford community, even, no especially, among those who are marginal. Most of us have been outsiders at one time or another.

Another way to view slacking is as a utopian anticipation. This is difficult for driven adults to accept. Overburdened kids who shut down move off the success track because they may value other routes, along which they find themselves— much as Shapiro did when he ran across the country. Increasingly, kids take a year off between high school and college, and many now finish college in five or more years. And they live at home or return home after college. Having watched their parents stress out, and struggle to pay their mortgages and credit card bills, kids may conceptualize success differently. They may value time with friends, community, volunteer work, music, writing, travel, self-exploration. Many of us during the sixties decamped to Europe over the summer in search of ourselves. We embarked on the equivalents of the transcontinental run, and many of us are still in search. We have learned that there is no finish line; we seem rooted, but the Earth continues to spin.

Adults slack, too. Only 40 percent of Americans work traditional nine-to-five hours. Some cannot find full-time employment. But others value part-time and shift work; they want time for themselves, hobbies, exercise, family. Feminists establish careers and then take years off in order to raise children. They home-school. Men join them. The fast track is not for everyone. Slowmodernity suits not only sullen slackers but many adults who want out of the rat race. They downsize—their homes, jobs, budget, expectations. Kids' freedom is contagious.

Perhaps the generations are at war because adults secretly resent and envy kids' freedom. Adults want children to join the pre-labor force, with all their daytime and even nighttime hours accounted for, because they want kids to share their pain. This is easily rationalized as getting ahead of the curve and a leg up—building a college resume and positioning young selves in the future labor market. But only a very small percentage of high-school graduates attend elite universities, the Yales and Wesleyans that have admission rates of 10 percent or lower. Most attend colleges and universities with nearly open admission policies. It matters little where one gets one's first degree; what matters is the last degree, and good public research universities, with affordable tuition, provide excellent platforms for careers. Adult pressure to get good grades and high test scores, and thus gain entrance into the "Ivies," is more about adult ego than children's need.

At issue, then, is the very definition of success. Fast-track baby boomers have children who will underachieve by choice. "Achievement" is not what it used to be. The personal cost in time and stress may not be worth the extra places in class rank or fractions of GPA points. These pressures on kids kill the love of learning, perhaps America's most pressing social problem. Slackers are those who choose to achieve differently, even in forms unrecognizable to their middle-class parents: web page design, video work, music, part-time employment with ample leisure time, hobbies. The payoffs: less stress and debt, more time.

Slacking is often expressive, taking literary, visual, and musical forms. Youthful slackers produce the iPhone discourse I discussed at the outset. They crave connection, but not necessarily with adults. Some belong to bands; others write poetry. In these senses, slacking has a creative impulse, but it rejects adult definitions and yardsticks, such as degrees earned, money made, residential square feet owned.

The boundary between success and failure is flimsy and debatable. Time may be the most precious resource for members of the anti-achievement generation—time to be oneself, to avoid evaluation, to be free of adult authority. Parents and teachers worry that their kids will not be able to support themselves. They also worry about the generations that will inherit the Earth. Rejection of achievement isn't achievement itself; it merely opens a debate about what counts. Slacking is decidedly pre-political. Most slackers are not schooled in early Marx and his humanistic conception of free work performed in sheltering community. It is not only that they aren't socialists. They are antipolitical because they view politics as yet another adult venue for deception and inauthenticity. In other words, slackers don't theorize their own alienation and their responses to it. They don't theorize at all.

This is because anti-achievement is formulated within the frame of reference of capitalist achievement—money, credentials, trophies. It is "anti," not "post" or "other." But the critique of capitalist materialism must be materialist itself; it cannot avoid power, wealth, domination. Slacking is pre-political where it portrays class struggle as kids against parents and teachers, ever the oedipal dynamic. But I am suggesting that kids are a social class, a pre-labor force dominated by adults who resent kids' freedom and want to prepare them for the alienated grind of adult life. Hence, adults reject the secret writing done by kids off the grid of academic convention. This writing does not count, according to the usual adult criteria of achievement. In other words, adults recognize the political threat posed by slacking even where their children do not. Slacking calls into question basic value systems promoted by capitalism, especially the value placed on workplace duty.

Kids were more political during the sixties, at least some kids. It is risky to paint with too broad a brush. Some sixties kids were conformist, preferring pep rallies to antiwar rallies. By the same token, not all anti-achievement-oriented kids today fail to theorize their own alienation. Some take political action by joining political movements, others work for non-profits, and still others embark on the long march through the institutions, transforming adult society from within by growing into and yet transforming adult roles in education, culture, business. Sixties kids had the example of Kennedy's New Frontier to support their idealism. Today, politics means corruption to the young. Irony replaces passion among cool young people. Not to commit is the primary commitment.

There is irony in an adult theorizing kids' resistance and rebellion as potentially political moments, even though slacking is the rejection of adult condescension. But the young aren't fully formed and need guidance, much as we were young once and could have used some sobering by elders sympathetic to our causes. Indeed, the New Left, both black and white, was not simply a youth movement but a generationally multilayered alliance of teenagers, kids in their twenties, and even oldsters beyond the ripe age of thirty! Dave Dellinger, a movement veteran, schooled younger antiwar protesters, much as Ella Baker helped the SNCC kids organize themselves with full awareness of the history of civil rights.

Not to listen to the kids, especially their secret writing, risks a generation gap so wide it cannot be breached. Disdain is not the correct posture, even though it may be difficult for boomer parents and elders to take slacking seriously as a pre-political posture. Few of my students are really motivated, and the empty chairs outnumber the chairs occupied by those who care enough to attend class, even if most of those endure college simply to get a credential and then a job. It is difficult for caring adults to value slacking positively when, to many of us, it seems quite anti-intellectual and, if anything, dystopian—simply retreating from the prevailing world.

Slacking, like earlier beat and hippie rejections of the status quo, can be redeemed as a utopian impulse if it understands itself as productive—of good works, culture, community, the self. Productivism is the opposite of this, simply valuing output for its own sake. To urge a slacker to become a productive member of society would be instantly off-putting, given the meanings of productivity and work today. But if slacking is read as protest and as a secret utopianism, then this redemption of slacking as prefiguring a new world is not far-fetched. It is a protest of authoritarianism, conformity, bureaucratic routine. And it suggests a world in which aimlessness is actually purposive—purposive purposelessness, as Kant said. The hippies wanted a certain kind of world, which they enacted when they sang, danced, poetized. This was theorized by Ken Kesey and Tom Wolfe, who understood the counterculture not as "against" culture but as an alternative to the parental cultures of war, aggression, discrimination, self-denial.

Perhaps slackers have in common with beats and hippies an aversion to the Protestant ethic and the Puritanism on which America was founded by white English colonizers. Anti-Puritanism isn't enough on which to base a revolution, but it is a start—a turning away from delayed gratification in service to capitalism. Marcuse has theorized the pleasure principle and play impulse as crucial to a new society in which people experience their work as play, nature as a friend, and other people as hospitable members of common humanity. He blended early Marx with Freud to prophesy a society in which basic repression, required to keep the person whole and able to have productive relations with others, would not be confused with surplus repression—repression of creative and libidinal impulses beyond what is required by a non-alienating civilization. He speculated that we could blur the boundary between work and play, creation and self-creation, to the point of near identity. Where we could not "erotize" certain unpleasant jobs, we would automate them, freeing up labor time for self-creative labor—praxis, borrowing early Marx's Greek term.

Marcuse, alone among the Frankfurt School theorists, was patient with the sixties New Left and counterculture, even though he recanted some of his earlier optimism in *Counterrevolution and Revolt*. In that book, Marcuse surveyed the wreckage of the sixties and decided that the Right crushed progressive movements but that progressives contributed to their own demise by eschewing theory—reason, by another name. Infantilism replaced reason, the effort to think things through and to plan, much as the child's id isn't sufficiently checked by the superego and runs amok. How else to explain the Weathermen, the sixties bombers who derived from SDS, and the more nihilist components of the counterculture partial to drugs? The late sixties were days of rage that included moments of ecstasy.

Young people during the sixties wrote passionate polemics and manifestos. Words animated those difficult and exciting times. But these were mainly political writings. Even the poetry of Lawrence Ferlinghetti, like the novels of Jack Kerouac, had political undertones and implications. It is more difficult to trace politics in the songs of Nirvana and rap lyrics. Personal blogs and social-networking postings are rarely political. We are awash in a culture of narcissism, as Lasch (1979) termed it. And yet the sixties were personal, too, as existential angst involved in growing up inflected the political passions of the white and black youngsters who participated in the antiwar and civil rights movements.

Perhaps the difference between then and now—between New Leftists and hippies on the one hand and slackers on the other—is that there are few movements today and that, in the wake of the sixties, politics has been largely discredited. Obama is also a product of the 1960s, as civil rights cleared the way for a biracial candidate to succeed in the twenty-first century. Jesse Jackson connects the two eras, now separated by almost fifty years. He wept on the balcony of the Memphis hotel in 1968 when his hero Martin Luther King was assassinated, and he wept again when Obama was elected in 2008. Tears of tragedy had become tears of joy. And yet I'm not sure that Obama turns young people on to politics. He was cool, especially by comparison to Bush, McCain, and Romney. His coolness was interpreted by the young not as passion but as a postmodern distancing from politics, perhaps the outcome of Obama's studied centrism—his ability to appeal to diverse constituencies in deftly decentered discourses designed to please all of them.

And yet. Just as one notices depoliticization among the young, one also notes MoveOn.org and the revival of the Students for a Democratic Society (SDS) and its counterpart Movement for a Democratic Society (for baby-boomer elders). Thefbomb.org is a feminist blog for teenage girls. Feministing.com and goodasyou.org are feminist and gay-rights sites that keep the embers of social change glowing, albeit in a fluid, edgy, amateur format that, like Montaigne, allows people to essay their own lives outside the boundaries of the official culture industry, which requires one to have an academic job or a literary agent.

In the next two chapters, I consider the theft of time, even of sleep. Time theft breeds time rebellion, even if youthful time rebels don't fully understand themselves as participants in a long tradition of protest against capitalist regimentation. They may simply accept adults' terms for them: slacker, loser. But there is something political going on as kids sleep in and thus engage in their own meaningful absenteeism. Later, I link time rebellion to schooling, which is a ground of conflict over children's freedom and discipline.

CHAPTER 8
TIME ROBBERS, TIME REBELS

Kids are chronically short of time, and they push back. They stay up late and sneak in moments of respite, which often include phoning, texting, messaging, posting, blogging. Kids are short of time because adults steal it from them. There is an economy of time, which is finite. It could be said that exploiting people can involve controlling their time use, the hours they otherwise have available to live a free life. All adolescents and their parents understand that the waking hours are challenging because everyone stayed up too late and school starts too early. There is too much to do, too little time to do it.

Adults rob kids of time and kids try to wrestle it back. They rebel against time robbery in these ways:

- They stay up late, even past their parents' bed times.
- They sleep in and struggle to get up for school.
- They postpone homework and other assignments until the last minute, if these get done at all.
- They defy school authority by skipping class, arriving late, and using in-class, hallway, lunchroom, and playground time for texting, messaging, chatting—against school rules.
- They engage in schoolwork slowdowns, dawdling over in-class work and tests.

- They violate the nighttime curfews imposed by their parents.
- They pretend to be doing schoolwork and homework when in fact they are texting.

Capitalist Time

On the one hand, this sounds like random mass disobedience, wanton adolescent behavior. On the other, it closely resembles workers' activities of resistance early in the union movement, in England, Europe, and the Americas. Workers would "work to the rule," doing no more than the minimum in order to protest managerial control and domination. Kids don't necessarily know that they are resisting, and they are not following the manuals provided by Karl Marx or the Webbs. Theirs is a protean, pre-political resistance—an instinctive, mimetic, spontaneous rebellion that flows as much from their tired bodies as their racing minds.

Marx in the nineteenth century felt that the ultimate capitalist contradiction would reveal itself in general unemployment, occasioned by dysfunctions built into the profit system that allow a few to become inordinately wealthy. Oligopolies and monopolies, emerging from an earlier market economy, would put smaller businesses out of work and throw their owners and workers into the Sargasso Sea of the industrial reserve army—the unemployed and underemployed. Today, this contradiction has not been eased; if anything, it has deepened as class struggle has been spread to the global system through what Marx termed "uneven development." This cleaves the world into wealthy and poor nations and provides breeding ground for dictatorship, religious fundamentalism, terror, starvation, and the spread of HIV.

Young and old are similarly cleaved and becoming more so. The economic contradiction of rich and poor, those who own capital and those who have only their labor to sell (assuming there is even a buyer of it), has been overlaid by a temporal contradiction, which also limits capital because there are only so many hours in a day. Young and old struggle over time—who will control it, how much sleep people get, how it is occupied. Young people want more time, and adults worry that they will misspend it, becoming slackers.

Marx well understood that capitalism operates through time. This is time in two senses: capitalism has a past, present, and future; and capitalism utilizes time in order to produce and reproduce itself. Marx's final book, *Capital*, brilliantly analyzes the role of the clock in exploiting workers. It is precisely through the time of the working day that capitalists have figured out how to extract what Marx called surplus value from workers. There is a time each workday during which workers are uncompensated for their labor; this is the time from which profit flows. There is also a possible time, ahead, in which people free themselves from managed and exploitative time and begin to experience the passage of time as unencumbered experience, what Henri Bergson called duration. As I explore below, free time can function as a utopian imagery, especially for our kids whose lives are strictly regimented.

Time helped Marx understand how the seemingly fair exchange of daily labor for a wage was in fact unjust. He theorized that the worker labors for a portion of the working day in order to transfer enough value to the commodity that, if the workday ended then, the sale of the commodity would produce enough value (exchange value in Marx's terms) that the owner of the company would recover his or her investment—in materials, plant, spoilage, advertising, transportation to market, overhead, and wages. Everyone would break even. But the bourgeois political economists insisted that capitalists must receive an additional return on their investment, moral compensation for taking the desperate risk of starting a business in the first place. Where does this "profit" come from?

It is tempting to view capitalists as Scrooges who dole out a meager shilling or two a day, purposely depressing wages so that they can live in opulence. But that would risk workers' resistance and revolt. It would also impoverish them so much that they could not consume the commodities they and other workers produce. Capitalism, as John Maynard Keynes well understood, needs workers to view themselves also as consumers, negotiating the balance between their public (work) and private (leisure/consumer) roles so that consumption can match production. Only if consumption matches production can capitalists retrieve from commodities otherwise languishing as inventory not only their original investments but also the portion called profit. Capitalism, in fact, as Henry Ford brilliantly understood, wants to keep profit margins relatively slender, relying on mass production to help amass large profits (Aronowitz 1992; Braverman 1974).

Indeed, mass production, to be matched by mass consumption, would have been impossible had Ford priced his Model T too high. He wanted to put work- ing- and middle-class Americans on wheels, allowing them to escape the din of the early-twentieth century city in order to picnic in state and national parks on the weekend. This assumes an inviolate wilderness carefully compartmentalized as "nature" and not part of urban sprawl—a tenuous assumption today. Indeed, Ford voluntarily doubled his workers' wages in order to enhance their ability to buy his cars and to engender loyalty to his company. He also gave them a two-day weekend, both in order to shop and to re-create themselves for the grueling stint of the work week. (Marx, too, advocated a five-day week, as Tom Hayden reminded me in our discussions about the legacy of early SDS.)

Marx and Ford identified two complementary parts of the same process of time theft. Marx identified the need for time to be regimented and controlled in the workplace so that workers would work without compensation for a portion of the working day. Ford identified the need to free people from work time so that they could spend money and restore themselves after the drudgery of factory labor. This is not to suggest that leisure is really free; it is not in a fast capitalism in which downtime is as severely regimented as work time. Many people watch television (see Miller 1988) or surf online after work. The television schedule is rigidly structured and displayed in newspapers and available online. If people cannot manage to find time for their favorite shows, they may tape them and watch them later—an apparent temporal flexibility betrayed by the compulsive need to keep up with the unfolding serial. DVR and TiVo manufacturing has become a major industry in response.

Marx stressed time theft in order to demonstrate the origin of profit in workers' uncompensated labor. Ford stressed the need for time management as workers were, in a somewhat later stage of production, able to enjoy a few more hours of apparently "free" time. Ford was advancing beyond Marx as capitalism in the 1920s and 1930s entered the era of mass production, which cheapened commodities such as the first mass-produced automobiles and inserted technology into the production process, especially with the advent of the assembly line, later to be followed by robots and computers (see Mandel 1978). Ford was working and strategizing nearly half a century after Marx. In the interim, people's rudimentary needs could be more readily satisfied, and they were encouraged to work and shop beyond those basic needs. As mass production evolved through both twentieth-century world wars, the primary challenge for capitalism was to persuade people to work and shop beyond the level of mere subsistence. As the Frankfurt School recognized, beginning in the 1930s, capitalism had to produce "false" needs, or wants, through advertising for the latest technologies, which were planned to be obsolescent. In this context, time became a more significant factor in capitalism, as people were manipulated not only in the workplace but in after-hours spheres of their existence and experience. The culture industries were designed precisely to bring about the total administration of time, blurring the boundary between the economy and entertainment.

For Marx, writing in the mid-nineteenth century, urban penury was so oppressive that he did not have to theorize consumption, for example distinguishing between needs and wants. People wanted bread, clothing, housing. The leisure class had not yet appeared on the horizon. For Ford, and for post–World War II theorists such as Veblen and the Frankfurt School, leisure time needed to be theorized and administered so that people shopped sufficiently and in ways that did not harm their work performance, such as overindulgence in alcohol (see Ewen 1976; Leiss 1976). Time was always a central issue in capitalism—for it is through unpaid time that capitalists derive profit from labor—in an era of potential affluence, at least for the majority of people in industrial countries. Time became even more important as capitalism needed to control not only work time but all time. It was in time away from work (or, now, at work using ICTs) that capitalists could ensure that people were kept busy shopping, both fueling the profit machine and remaining distracted from larger questions.

Total Administration of Time

Ford couldn't be certain that his workers would handle their doubled wages and an extra day of weekend time responsibly. In a stunning and prescient move—anticipating the total administration of time more than half a century later—Ford created a "sociological department" staffed not by degreed academic sociologists but by people who functioned as social workers, truant officers, and union busters to supervise workers' use of time and money. Members of the sociological department went door to door and canvassed Ford workers to make sure that they were not absent from work and were not drinking away their wages. The sociological department also forcefully resisted Walter Reuther's attempts, during the 1930s, to unionize auto workers. Well before Foucault, Ford understood that people needed to be supervised (or believe that they were being supervised) so that they "spend" (in both senses of the term) their leisure responsibly.

There is a telling dialectic at play here: as people enjoy shorter work hours, they must be encouraged to use those hours in ways that benefit capital. Hence, we watch television and surf the Internet, bombarded by advertising images of the "good" life. Yet the intensification of time robbery and time administration robs people of the time and space in which to re-create themselves, reproducing their selfhood and identity in ways that allow them to function during "public" time.

Where Adam Smith preached abstinence and thrift in the mid-eighteenth century, by the twentieth century Keynes recognized that capitalism could only thrive if people and government spend beyond their means. Saving was to be replaced by shopping, a process rapidly accelerated after World War II with the advent of personal credit (see Packard 1959). Endless shopping was to match endless production, a process first understood by Marx, who analyzed workers simultaneously as consumers. To achieve endless consumption, post–World War II capitalism, using information and communication technologies such as the telephone, fax machine, cell phone, and now the Internet, enabled round-the-clock spending. This was accompanied by a blurring of the boundary, heretofore inviolable in modernity, between the public and private spheres. The Frankfurt School recognized that privacy was imperiled by what I am calling the total administration of time or fast capitalism. The problem is that the total administration of time breeds massive discontent and psychic rebellion—alienation, to use an earlier vernacular.

Early twentieth-century environmentalism assumed the distinction between the unsavory urban life and unsullied nature. With the malling and suburbanization of America, that image is a thing of the past. Instead of driving to nature, people have to fly to it, if "it" is to be found anywhere at all. The automobile has become not a means of escape but a home and an office. Women rushing to work in the morning apply their makeup while stopped at the red light. People of both sexes talk to the office and to their kids on their cell phones while driving. People drive to work, as they did before, but now they also drive everywhere else. Americans walk much less than their European counterparts, which may explain why obesity is a more serious problem for Americans. The era of Walter Benjamin's (1999) *flaneur*, walking unhurriedly and with curiosity around great cities, has long passed, at least in America.

Part of this story about the management of leisure as well as work time is also, and implicitly, a story about space. Critical theory, with a few notable exceptions (e.g., Adam 1995, 1998; Harvey 1989; Hassan 2003, 2012; Soja 1989), has neglected both space and time. Time has become compressed in fast capitalism largely because things are so far apart and the only way to get to them is by driving over urban, regional, and interstate roadways. And the malling of suburbs, coupled with the decline of urban cores, has usurped even more time for interurban commuters and shoppers. Jane Holtz Kay (1997) tells this story nicely.

Ford was such an important figure, if not a trained social theorist, because he understood the connections among the automobile, cities, nature, and time use.

Ray Kroc and Walt Disney, World War I compatriots, had a similar sophisticated understanding. The era of mass production, which involved the administration of work time and space on the assembly line, cheapened cars and made them available to workers who formerly went to work on foot or by streetcar. Although initially people went on picnics away from the city to escape their urban alienation, the road less traveled was quickly overrun by streams of weekend commuters. Think of the exodus to and from Long Island on a New York City summer weekend.

Initially, the automobile helped people escape their alienated urban lives. And weekends could be spent re-creating the self and the selves of one's family members. But gradually this time became compressed (not enough of it) and administered (managed by stewards of the culture industries). Images abound from the post–World War II period in America of people sitting in highway gridlock, privatized, their vehicles guzzling gas and polluting the atmosphere. They are going nowhere fast. The REM music video "Everybody Hurts" speaks to this post-urban alienation, which was certainly not intended by Ford. This video captures the lived experience of a post-Fordist, Los Angeles–like sprawl in which time and space have broken down, connoting the imagined experience of the aftermath of a nuclear holocaust in which both cities and norms dissolve (see Davis 1990). Ford's utopia of the five-day work week, teetotaling families, and picnics on the weekend has become a dystopia of compressed and managed time and gridlocked, polluted space, disrupting capitalism in ways unimagined by Marx, whose depiction of alienation might have been understated.

Horkheimer and Adorno argue that Marx's critique of capitalism was not sweeping enough. Indeed, they argue that civilization began to erode as early as ancient Greece, whose philosophers attempted to master nature without considering the repercussions for humanity. What the Frankfurt theorists call "domination," borrowing a Weberian term, is actually more encompassing than Marx's "alienation," referring to all Promethean projects that seek total control as a way of validating the externalizing, conquering self. They would have readily theorized the total administration of time in this way, noticing that the compression and acceleration of time produces a dialectical backlash, much as the domination of nature triggers nature's revolt. This revolt of nature is captured in cultural works all the way from Rachel Carson's *Silent Spring* (1962) to blockbuster movies like *The Day After Tomorrow*, in which the world is beset by meteorological catastrophes caused by global warming.

Truth can be stranger than fiction. The 2005 tsunami that pounded Asian coastlines is a stunning testament to nature's apparent malevolent agency, as was

Hurricane Sandy. And the destruction of the World Trade towers could have been depicted in a Hollywood horror movie starring Bruce Willis. The total administration of time belongs to the same dialectic of domination and disruption, stemming from the Greek effort to conquer nature and having matured in a cybercapitalism (Dyer-Witheford 1999) that makes global instantaneity possible.

Reproduction of Selves Thwarts Reproduction of Capital: A Port Huronized Slowmodernist Critical Theory

The total administration of time saved capitalism from itself, given its internally contradictory nature, first recognized by Marx. But it also threatens to thwart capital formation and accumulation as the administration, colonization, and intensification of time runs up against its outer limits: the working and shopping day cannot be expanded beyond 24/7, given people's needs for sleep, recuperation, recovery. The reproduction of the self slows and stymies the reproduction of capital and creates a fertile ground for rebellious, transforming projects and movements. Capitalism cannot afford to let people rest, and it cannot afford not to.

Time robbery breeds time rebels, who could become political rebels once we theorize time as an essential political category in fast capitalism. Time robbery deprives people of sleep, affect, the vestiges of autonomy with which to theorize and strategize about their system-serving use of time. Time robbery, when taken to its limit, robs the object—fast capital—of a subject, the self. Eventually, the system will shut down as people reach overload, deprived of time, family, leisure, nature, self, identity with which to make informed choices. Even if they don't become rebels and revolutionaries, they cease to function productively. The only means available is to medicate them with alcohol, speed, anxiety drugs. It is not clear that this is a viable long-term solution for time robbery, as Huxley (1953) well understood.

Marx explained the basic contradiction underlying or, perhaps better, inhabiting capital as the fateful conflict between capital and labor. The former produces value for the latter, which as a noun refers to both a class and to amassed productive wealth. Labor expends itself through time, which becomes a means of depriving labor of a portion of value, thus allowing for private profit. Labor is not allowed to either possess or manage time during the working day. Now, in fast capitalism, time management extends beyond the job site into private downtime and even into nighttime. Influenced by Hegel, early Marx argued that the protean human project is work, which allows us to externalize ourselves in nature, fulfilling our desire to make, create, produce, build. I wonder if a better way to say this now is to argue that the fundamental human project is to have and use time freely, unbound from clocks, schedules, supervision, and now the rapid information and communication technologies tethering us to work, entertainment, and shopping.

Building on Petrini's image of slow food as a metaphor of slow life, we can conceptualize utopia not as the fulfillment of modernity but as a "slowmodernity" in which we blend modern and pre-modern in a dialectical synthesis that forgoes neither useful (Schumacher's [1973] appropriate) technologies nor the social relations and pleasures of pre-industrial Gemeinschaft, such as Petrini's slow dining. This is not Luddite per se, nor antiscience. It is a version of Nietzschean/Freudian critical theory that chooses everyday lives no longer governed by capitalist principles of productivity and performativity. At issue in a critical theory reformulated as a theory of time is the idea of *free time* as an equivalent of social and political freedom, a latter-day version of early Marx's concept of disalienation.

Writing in the mid-nineteenth century, Marx could not foresee the welfare state, culture industries, post-Fordism, or the Internet; thus he did not imagine how administration could become total. This is the point made by both the Frankfurt School and Foucault. Neither Adorno nor Foucault conceived of an exit strategy for selves whose lives, especially the lives of children, were accelerated and administered, now nearly 24/7. The Frankfurt theorists, in books such as Marcuse's *One-Dimensional Man* (1964), Adorno's *Negative Dialectics* (1973a), and Horkheimer's *Eclipse of Reason* (1974), emphasized the negative tendencies of the total administration of life in late capitalism. I include time robbery on their list of usurpations of humanity and freedom. With the possible exception of Marcuse, especially in his 1960s writings, the Frankfurt theorists did not also emphasize the dialectical—potentially positive—implications of total administration, such as the implosion and possible reconstruction of the social system and everyday life. They had a particularly pessimistic view of capitalism after the Holocaust, and with good reason.

Marcuse concludes *One-Dimensional Man* with comments on what he calls "the chance of the alternatives" (1964, 203–248). In 1969 he published *An Essay on Liberation*, which earlier had the working title "Beyond One-Dimensional Man," stressing the emancipatory possibilities opened during the social movements of the 1960s. None of the Frankfurt theorists emphasized time robbery— or, perhaps better said, time fascism—as an element of one-dimensionality and total administration, although their analyses of the closure of the universe of political discourse could certainly include my discussion of time.

Here I want to stress the positive potentials opened by time robbery carried to such an extreme that time compresses into sheer nothingness, preventing any light from shining through. These are the moments when people feel crushed by their accumulating and sometimes conflicting obligations, roles, schedules, stimuli. Although the effects on the self are always damaging, people can be turned into rebels as they theorize—think through—why this is happening to them and what they can do about it. This is dealing with domination at the level of the lifeworld, of everyday life (*Lebenswelt*, Husserl's term). There is another possible positive outcome of time fascism: people simply cease to function as diligent workers, consumers, and citizens at the high level necessary for fast capitalism to thrive. As people resist the total administration of time, they enter a timeless zone of retreat and reflection in which they shut down their electronic prostheses, put away their credit cards, and cease working overtime anytime, anywhere. Even if they don't fully understand what is happening to them, and what their options are, retreat from total administration is equivalent to, in Mario Savio's New Left terms, placing one's body on the apparatus, slowing it down.

In these seemingly post-Marxist, postmodern times, it is very difficult to convince people that socialism and liberated nature have any meaning for them. Neoconservatives and postmodernists meet on the ground of their rejection of Marxism, even if their argumentation is different. I am suggesting that a reformulation of early Marx's utopian goals—slowmodernity instead of socialism, even if their meaning is nearly identical—would help invigorate the leftist project. Other metaphors can be marshaled as well. The American New Left argued for "participatory democracy" as a ground-level device for countering bureaucratic hierarchy, of both the corporate capitalist and state socialist kinds. Indeed, the SDS's critique of Cold War era capitalism was forged out of Tom Hayden's and Dick Flack's readings of early Marx and of C. Wright Mills, both of whom would probably have embraced a critique of time robbery had they paused to theorize the relationship between clocks and capitalism.

A *Port Huron*–like version of slowmodernist critical theory—risking a jumble of metaphors and slogans!—is especially apposite in these times of American military adventurism, right-wing evangelical Christianity, gay bashing, and the othering of foreigners. The first decade of the twenty-first century felt like the 1960s, especially with Bush's reelection and the quagmires of Iraq and Afghanistan, which cost the lives of over six thousand American troops and $1.4 trillion, not to mention galvanizing the Arab world in potential jihad. We will have to await the completion of Obama's second term as president to determine whether he represented a fundamental alternative. Although critical theorists and postmodernists alike abandoned talk of the self or subject as inadequate in an era of late-modernist gigantism, not to mention talk of a collective subject, the times are right for return to a vocabulary of new life and new social structure in the gestures of a gentle everyday life. Time would be central to this agenda. A "next" Left would retain from early Marx and Marcuse the urgency of disalienation in the ground of everyday life, and it would speak a somewhat new vocabulary of slowmodernity and participatory democracy, playing on people's experiences of accelerated daily life and their feeling of utter powerlessness in their everyday worlds.

All of this is possible because ICTs, especially the Internet, have changed our world in significant ways since the late 1960s, when Marcuse was temporarily optimistic about the New Left. The world is now global, even if Marx had already anticipated globalization as an outcome of the logic of capital. But it has become global via mechanisms of instantaneity unimaginable to theorists of the nineteenth century and early twentieth century such as Marx, Durkheim, and Weber and even to those of the mid-twentieth century, when Frankfurt theory was composed. Instruments of instantaneity such as the cell phone and the Internet compress time and make the administration of time relentless and inescapable.

Marx anticipated globalization as he suggested that the "logic of capital" causes capital to ignore national boundaries and colonize the planet wherever it can find fertile ground for production, consumption, and resource extraction. One doesn't even need Lenin's concept of imperialism to explain the near-identity of Americanization and globalization, given Marx's own analysis of the compulsive, invasive tendencies of capital (see Hardt and Negri 2000, 2005). Habermas refers to all this as the colonization of the lifeworld, blending Marx, Durkheim, Weber, and Parsons. He argues that we must protect everyday life from imposed meanings, strictures, and structures. Time is certainly in play as ICTs allow people's uses of time to be bent to the purposes of capital, bureaucracy, and nation, colonizing people's circadian rhythms.

Time Rebels: What People Can Do to Reclaim Time and Their Lives

Theory confronts practice on the grounds of everyday lives. Phenomenological and existential leftists such as Paci (1972) and Sartre (1976) converge with Freudian "new" leftists in agreeing that change is worthless, and probably counterproductive, if it doesn't pass through the self. The model of a long road to radical change is ultimately endless, sanctioning short-run sacrifices of liberty and life themselves. The theft of time, and its total administration in a fast stage of capitalism, can be matched and bested by time rebels, people like you and me who think through these issues and decide to live differently, prefiguring and working toward a different world.

In *Speeding Up Fast Capitalism* (2004a) I discuss ways of slowing it all down; my final chapter is entitled "Slowmodernity." I offer not only desiderata but accounts of what people are already doing to reclaim time and thus their lives. As I argue there, this does not in itself transform capitalism but is a beginning, a necessary way for people to carve out space and time to become thoughtful and caring. These choices are aspects of what Marcuse termed the Great Refusal, the decision of protean selves, illuminated by theory, to reject the quotidian and to live differently. In the 1960s these choices often resulted in communal and quite pre-industrial living arrangements, even in cities. Action provoked reaction and repression. The Berkeley police action against the architects and occupants of People's Park in 1969 refracts conflict over the proper uses of the public sphere and nature. The New Left ran up against the repressive machine of Nixon's White House and the FBI, which launched an organized effort labeled COINTELPRO to rid America of young leftists. The Right "won" the 1960s politically by launching a successful and long-lasting counterrevolution—first Nixon, then Reagan, then George W. Bush.

This is not to suggest that the 1960s were a political failure. Young people and even some of their parents stopped the war in Vietnam, won important civil rights for minorities, and initiated the second wave of American feminism. What we learn from the 1960s is that change must begin at home, even if it cannot end there. The problem with orthodox Marxism, and especially the Soviet variant (Marcuse 1958), is that the long road to socialism never seems to end. And it is a road whose route is dictated by elites who never wither away.

So kids and adults alike can become "time rebels" and use this important wedge issue to bring about social change. The issues in the 1960s were civil rights and Vietnam, and today it is time dominion. We experience the administration of time in our daily lives, much as black people and their white allies were radicalized by the Klan and draft opponents and resisters were galvanized by the threat of dying in Vietnam. This experience of being crushed by time pressure can be a politicizing one. It can lead to time rebellion, auguring other rebellions, all designed to change the entire social and economic system.

Time rebellion avoids Luddism. My vision of slowmodernity blends anti- and pro-technology postures in a twenty-first-century synthesis. We can use the Internet instead of being used by it. We can continue to drive cars, but hybrid and sustainable ones. We can further automate industrial production. But we need to break the clocks—a metaphor—in order to purposely slow down our lives as a way of becoming more human and of effecting a prefiguring radical change. By prefiguration I am returning to a Freudian image of everyday people making different and better choices, armed with a Hegelian-Marxist "reason" (see Marcuse 1960). Marcuse makes it clear in his study of Freud and Marx that we need not abandon high technology as long as we use it for human purposes, thus fundamentally changing its logic—"pacifying" it or perhaps in today's parlance "greening" it. He borrows from Schiller and Nietzsche in positing a gay or new science and technology that would play with nature and with concepts instead of dominating them, draining them of life.

Marcuse suggests that false needs perpetuate domination far beyond what is necessary for survival. By keeping people shopping insatiably, advertising immobilizes them politically. Really, people immobilize themselves; they have choices, even of utopia. Although Marcuse doesn't talk much about time, acquiescence to time robbery is clearly a false need in the sense that people feel unnecessarily pressured and thus pressure themselves to spend inadequate time—and therefore freedom—in reflection, enjoyment, noninstrumental activities. Clearly, early Marx and Marcuse would include time among the most important political variables and values. And he would suggest possible "times" that break with domination and do not succumb to the Weberian rational calculations at the heart of capitalist logic. In this sense, I believe that Marcuse would have endorsed the notion of slowmodernity as a blend of pre-modern and modern that, in dialectical synthesis, takes them to a higher level and thus provides a much-needed utopian imagery just when utopia seems a remote possibility.

Time, therefore, needs to be at the center of critical theories that attempt to understand people's servitude today, much of which is self-imposed. This is not to blame people for rushing around and courting the dangers associated with high blood pressure but to recognize that people must be involved in their own liberation, making better choices that prefigure utopia for all.

People, especially the young, already resist and refuse the theft of time and hence their lives. They must configure their time differently, where possible. Although singular heroic agency cannot change the world by itself, without people exercising agency nothing will change and time will only accelerate and compress. It will accelerate until people can no longer function in their waking lives but implode, simply unable to handle so much stimulation and deal with so little sleep and downtime. The ADD epidemic reflects implosion among the young. In the meantime, people can shut down electronic prostheses, such as laptops, cells and smartphones, and television, all of which dictate both pace and content. They can throw away their calendars and planners, turn off their alarm clocks, and banish clocks and watches when possible, refusing (and thus reconfiguring) the "timescapes" of an accelerated capitalism that equate time and money, punctuality and duty.

Sexuality is at once irreducibly intimate and transparently political, providing a medium of resistance and liberation. Time is equally intimate and political, both private and public. Without time management, capitalism couldn't function. But the self requires time off the clock in order to restore itself, both asleep and awake. In fast capitalism there is no time off the clock; all time is sucked up and made performative, productive, and reproductive. Advertising is a perfect example: people watch television to escape, but, while watching television, they may be stimulated by shopping networks to turn off the TV and shop online or by phone. If they call "just in time," they will receive a discount, which might be applied to the next purchase. Twenty-first-century sensibilities will experience, organize, count, and theorize time differently, unhinging time from production and performance and choosing to live unencumbered by imposed schedules and acceleration that compress and crush.

Time is thus an essential component of potential political agency, as the original existentialists understood. Bergson captures this well in his concept of time as *durée*, duration, not as a functional, divisible unit. But to experience time this way, as a moment of everyday sensibility that neither compresses nor crushes, is difficult, if not impossible, under a regime of time dominion. It could be said that an agenda of time liberation, which is central to a contemporary critical theory dealing with globality and instantaneity, is a reformulation of early Marx's communal utopia and of a rationality of gratification. We must incorporate time, just as we incorporate space, into the emancipatory project, especially because time, like sexuality, is such an important medium of domination today.

My concluding observation is that time robbery cannot continue without running up against the limit of sunrise and sunset, the solstices, growing old—sheer scarcity. The problem with capitalism, now as before, is that its logic is colonizing, refusing limit. But as Heidegger (2000 [1927]) recognized, our time is limited, and people naturally resist and refuse overscheduled, sleep-deprived, anxiety-ridden existence as symptoms of the damaged life. People escape, resist, shut down; for them—us—time becomes political. We simply don't have enough time, affect, money (always money) to soak up the commodities that our mechanical prostheses ceaselessly produce. We require sleep; we need to unplug. For capital not to recognize this, nor resolve it, is the real limit to capital.

Dan Wood (2013) tracks this limit to capital in his discussion of "powering down," the attempt of people on vacation to turn off their phones and laptops and to avoid schedules and programs. This inverts anytime/anywhere and suggests a utopian program of never/nowhere, although, I would add, it is inadequate to compartmentalize utopia as vacation time. Time rebellion intends to be always/everywhere.

These limits to capital are especially acutely felt among adolescents and children, who need sleep and downtime in order to grow and thrive. We have inadvertently returned to feudal concepts of children as miniature adults. Kids push back; they haunt the night, which is their world. They go underground, creating a secret society held together by texts and Facebook. They sleep in and stay out late. These are political acts, even if they are not intended to be. They are protests of fatigued bodies and wired minds.

CHAPTER 9
TIME WARS AND NIGHT MOVES

Limits to Capital and Sleep

The twenty-four-hour day imposes certain inexorable limits on capital and profit. Without sleep, workers cannot work effectively and shoppers run out of energy. Everyone needs to restore oneself both physically and emotionally. And yet, as I noted earlier, many adult Americans have jobs with nontraditional hours, part-time, nighttime, flextime, weekend work. And millions of young people inhabit the night as the only realm in which they are free of adult supervision and expectations. Time wars are conflicts over the administration and use of time. As such, they are partly generational conflicts. They are also class conflicts, with capitalists, now as earlier, requiring workers to work until they drop and shoppers to shop until they drop. Capital, as ever, has a temporal dimension. As I explored in the preceding chapter, it is only through the exploitation of the worker's free time (i.e., unpaid time) during the working day that sufficient surplus value can be transferred to the commodity so that profit would result.

Time war involves the battle over downtime, and especially sleep. A laptop capitalism needs business to be conducted around the clock, around the globe. The circuit, as Marx termed it, between production and consumption must be closed. Busy work must be matched by busy shopping, which becomes a kind of work. There is no nighttime; someone is online somewhere. Work, shopping, entertainment, and interpersonal connection merge and blur as boundaries such as day/night ease in a global capitalism.

A late capitalism, as Lukacs recognized as early as 1923, needs social relations and even consciousness to be "reified"—turned into things experienced as unalterable. Marx in volume one of *Capital* already discussed commodity fetishism as the tendency of work relationships to become ossified and object-like, impervious to human intervention. In a post-Fordist, laptop/Internet era of capitalism, reification involves a cultural ether or ambience that banishes talk of utopia and radical social change and instead immerses people in a certain "everydayness," precisely what Marcuse termed "one-dimensionality." In effect, culture must work overtime to divert people from the prospect of their liberation.

And so, heretofore "private" time, outside of work, needed to be colonized by advertising so that people would not only perform extra work and shop but would also remain oblivious to the possibility that life could be lived differently. They needed to be entertained, which has a commercial element. Our media culture is partly business and partly pleasure or, as Marcuse called it, "repressive desublimation."

An entertainment culture diverts people, mollifies them, keeps them awake. Watching television, surfing the web, sending text messages for long hours require chemical stimulants such as coffee and energy drinks in order to keep people awake when they might be sleeping. This only compounds the problem of sleeplessness, which I contend is one of the most important manifestations of what Marx originally called "alienation." Another and related manifestation of alienation is stress, and yet another is inactivity, which has obvious health and emotional consequences. Many deal with these manifestations pharmacologically, with mood-modifying drugs, blood-pressure medication, diet aids.

Sleep is an important limit to capital, as are mental and physical well-being. Sleep deprivation and disruption contribute to emotional and physical problems, as well as reflect them. Although flexible, time has objective limits. The work force and buying public (one and the same) need a measure of REM sleep in order to function. Sleep is compromised for the following reasons, all of which can be traced to a fast capitalism:

- People are tethered electronically to their work, which they carry with them and cannot easily escape. This invades the night.
- Shopping fills evening hours, especially where people can shop electronically.

- Entertainment, necessary to divert and restore people, occupies nighttime hours.
- To stay awake, people use stimulants, making it difficult for them to fall asleep.
- People, especially the young, haunt the night as they evade supervision and surveillance. People seek to be off the clock.
- Physical inactivity in a sedentary society in which nearly everyone drives and works sitting down makes it difficult to fall asleep.

Insomnia: Alienation or Resistance?

High blood pressure, stress, depression, attention-deficit disorder, and obesity are all health problems of an affluent society with too many consumer options, fast food, high-fructose corn syrup. Add insomnia to the list of alienations. I theorize insomnia as both a product of overstimulation and a reasonable response to the "total administration" of people's time. It may even be a utopian yearning—a time off the clock.

Bodies in motion, stressed and fatigued, need downtime. They require deep sleep, relaxation, repose, exercise viewed as play. Inadequate sleep, not enough or insufficiently deep, is a symptom of workaday alienation. And it reproduces itself, creating deficits that cannot be readily remedied by sleeping in on the weekends. Some people need less than others. Creative types may need less, or perhaps they nap, leaving night for their bursts of activity. A symptom of bipolarity may be the manic inability to sleep, which can stretch over days. And depression may be the name we give to the other phase, when stimulation wanes.

The medicalization and psychiatrization of social and human problems are useful in that we acknowledge the connection between mind and body; but they can be harmful where we imply that such issues are out of people's control unless they resort to drugs or therapy. The very concept of a "mood disorder," of which bipolarity is a prominent example, was challenged by the 1960s antipsychiatry movement, which questioned where "normal" leaves off and "abnormal" begins. Indeed, it is tempting to reverse that hierarchy and perhaps to view certain per-sonality disorders as blessed or creative states in comparison with the narcotized ever-the-sameness of most people's daily lives. Perhaps certain types and episodes of so-called mental illness are protest and transcendence, and psychiatry merely a means of social control and conformity.

Nonetheless, bodies need sleep and rest, just as they need motion. It is a striking insight that much malaise, even so-called mood disorders, can be tamed by vigorous exercise, as people produce neurotransmitters that succeed in stabilizing and elevating their moods where drugs have failed. And there are other obvious health benefits to be gained from walking, running, swimming, biking. It has been shown that exercise does better than drugs in managing people's moods (Carek, Laibstain, and Carek 2011). Those of us who run can attest that there are few better means and modes of disalienation than running our daily hour—a time off the clock during which we defy "administration" and the ensemble of productivist and consumerist expectations inundating us from waking to sleep.

I waver between viewing the deep night (or for that matter very early morning, before the sun rises) as a site of alienation or utopian yearnings. Perhaps it is both, contradictorily. The inability to sleep could be evidence that people are not getting their needs met during the day, when they perform alienated labor and so-called leisure. It could also be that they are overstimulated, if also undersatisfied, by their daytime lives.

There is not necessarily a clear boundary between day and night existence any more than there is a clear boundary these days between work and leisure. Both can be done anytime/anywhere, with the new informatic tethers and networks. By the same token, day bleeds into night as people stay up so late that they see the sun rise—a not uncommon existential condition among teenagers and college students.

Sleep might be the only real downtime, when people repair wounds, both psychic and physical, and where they dream the dreams of a new life and world. Freud, in a functionalist argument, suggested that dreaming is useful because it allows us to fantasize in ways that would be unhealthy in waking life. These could be dreams of death or life, Thanatos or Eros. Freud didn't anticipate the reduction of REM sleep, when the real dreaming takes place, because insomnia was not yet a general human affliction, requiring caffeine, alarm clocks, naps, and just sheer waking exhaustion. Nor did he anticipate a world in which boundaries between work and play, and day and night, would be blurred to the point of identity. In nineteenth-century Vienna, there was no round-the-clock cable television nor twenty-four-hour fast-food franchises. And there was no Internet, through which people inhabit the night together.

I hesitate to inflate insomnia into political resistance. Perhaps it is resistance, of a pre-political kind, much as young people's texting, tweeting, messaging, and blogging is a first stab at creating a new body politic, a drama-free and adult-free kidsworld. There may be a thin and permeable boundary between alienation and resistance; call it *escape*, the unlocking of the iron cage's door out of which the young flee screaming and singing. Alienation/escape/resistance is incomplete without the final stage—reconstruction, the making of a new world with vision and a plan. One must notice that visions often turn into hallucinations, or begin with them. Hallucination may be another word for, or type of, dreaming, as well as a drug-induced haze through which the world is softened and humanized. At least that was the posture of sixties hippies who used marijuana, LSD, and other drugs to amplify experience and sharpen perception.

One can get detoured, as most people are, most of the time. There is

- Alienation: feeling (and being) out of control of one's world.
- Escape: finding temporary relief, especially during the night.
- Resistance: putting boundaries around one's experience and one's world to protect oneself and one's group.
- Reconstruction: figuring out the causes of alienation and working to eliminate them and build a better world for all.

For most, especially kids but also their parents, the world feels crushing. Relief is sought—in shopping, entertainment, travel, the deep night. The search for relief often involves others as resistance mounts and people attempt to compartmentalize a world apart, time alone and together. The young, almost by definition, don't write or read manifestos. At most, they may evoke their desperate experiences in writing and art. Yesterday's *Catcher in the Rye* is today's blog or tweet. One writes oneself and one's battering by the world, reaching out to readers who would become writers. Call these readers friends, although today friendship can be global as time and place are unhinged.

Night Moves: The Young Take Back the Night

And so the young claim the nighttime and blogosphere as their own. They have little else in the way of solid identity. The young stay up late precisely because it is only then that they can evade adults. As parents slumber, kids form cybercommunities. Admittedly, these are alienated activities, not necessarily good for kids who need to restore themselves. But they are also instances of resistance and can be elevated into a political movement.

Why has nighttime become a political battleground, and especially a generational one? It is nearly the only realm in which adults do not or cannot engage in "total administration" of youngsters. Some parents try, imposing bedtimes and curfews. But the young are nocturnal, even if they may require more sleep than adults. The central issue here is the conflict and contrast between administration and freedom—control and agency. Adults seek to standardize and regulate the lives of the young as they prepare them (perhaps also secretly envying them) for alienated adult lives—the lives parents and teachers lead. This regulation takes several forms:

- Enforcing bedtimes and curfews and ensuring that kids awaken in a timely way.
- Enforcing school dress codes that address clothing, hair, piercings, tattoos, where on the waist pants fall.
- Imposing time-consuming homework, beyond what is necessary to add intellectual value to daytime schooling.
- Confusingly urging kids to abstain from sex while they also may oppose abortion. Condoms and birth-control pills would reduce the rate of unwanted pregnancy.

Kids push back on all fronts:

- They haunt the night, messaging, blogging, posting—forming like-minded communities.
- They break dress and comportment codes, often simply because, like Mt. Everest, they are *there*. School boards and school administrators across America believe that students would toe the line on all of these other issues if only they wore pants without holes in them, belts, and shirts with collars.
- They neglect to do homework or turn it in late, risking grades of "zero" on each of these deliverables.
- They have sex, sometimes unprotected. And they seek abortions.

It is tempting to view all of these acts of rebellion and resistance as "what kids always have done and will do." Freud and many others have noted that kids need to rebel, individuating themselves, learning how to deal with the conflict between impulses and conscience, establishing their identities. Freud argued that people, to join the social contract, must engage in repression, delaying gratification and engaging in sublimation (where they seek acceptable outlets for narcissistic and antisocial urges). Marcuse argues that in late capitalism, people engage in "surplus" repression—repression beyond what is really required for them to behave dutifully in work and other spheres of adult life. Technological abundance promises to relieve people from these heavy obligations, and thus capitalism must find ways to *deepen* alienation, distracting people from their possible freedom. Thus, Marcuse and his Frankfurt School colleagues extended the concept of alienation, as Marx termed it, and called it "domination." The means of achieving this domination are what they term "total administration"—the regulation of every waking and even nighttime hour.

And so one must question the extent to which kids need to be kept busy and taught to be diligent, so much so that they dress and act alike, go to bed on time, do mountainous homework, and refrain from sex just as they avoid birth control and condom use. Freud is certainly correct that we cannot let the id (a component of which is libido) run amok; we must learn to delay the gratification of our impulses and to channel them in nonnarcissistic ways. In terms of the discussion here, kids and adults, too, need sleep; insomnia and all-nighters are neither viable political nor personal strategies.

Blending Pre-Modern and Modern: "Slowmodernity"

People in primitive and agrarian pre-modern societies were not insomniac. Nor were they locked in time wars. The absence of artificial illumination meant that people shut down when the sun set, and they were physically tired from the hunt and the harvest of crops. Nighttime was not an alienated sphere of existence but a time for deep sleep. But this was set against a desperate struggle to survive. Only with the mass production of food and other accompaniments of modernity such as public health and public education in the context of the establishment and growth of cities could people rise above subsistence. But they still performed alienated labor and, eventually, even their downtime after work was colonized, producing the alienations of the night that I have been discussing.

We need to blend the pre-modern and modern, restorative sleep and exercise, with mass production and the end of hunger. I have characterized this dialectical synthesis as "slowmodernity" as I considered a fast capitalism sped up since the advent of the Internet. A future slowmodernity, combining the best features of the pre-modern and modern, would return us to pre-modern rhythms of time, work, harvest, leaving the deep night as the time for dreams and restoration.

Youthful alienation is especially acute in the context of schooling, the subject of the following chapter. Most kids hate school, which they experience as a prison. Schooling is closely linked to sleep deprivation and kids' nighttime occupancy of a counterworld apart from adults. I conclude with thoughts about how to make school a positive and meaningful experience.

Chapter 10
"I Hate School"

Slackers are young people who, above all, loathe school. They hate the structure, the endless homework, the compulsive testing and grading, authoritarianism, physical confinement behind desks, lack of free time, dress and comportment codes. This is primarily not the fault of teachers, although some, as alienated as their students, can be mean and unforgiving. Hating school has several causes. It's an outcome of a society in which kids constitute a pre-labor force being prepared for the real labor force. It is also a product of adult envy; we want to turn kids into ourselves, anomic occupants of office cubicles doing ever-the-same work, with little downtime. Hating school results from a conception of education as the ingestion and repetition of facts, a fill-in-the-blank mentality. Education is not viewed as a generative process whereby kids create themselves by coloring and writing outside the lines. Hating school is an outcome of assigning everything a grade. This leads to low self-esteem, which may be the most important manifestation of what Marx termed alienation. Finally, kids hate school because their minds and bodies are split; they do not get enough physical activity during the school day to allow the endorphins to flow. In short, school is not playful, nor is it creative, nor does it reward outside-the-box thinkers. It rewards drones and conformists. And virtually every child in America feels this way. That is the real pity.

The Twenty-Four-Hour Day

The Absence of Downtime and Creeping Homework

Kids lack the time to be themselves. Their school days are long and very structured, and they have to leave home early in the morning to start the school day. They often return home in the dark, especially during winter hours. Once home, their school day continues. Adults who work a forty-hour week begin their down time, at least theoretically, when they return home. In laptop capitalism, their work assignments might still pursue them, what with email and messages to answer and deadlines to meet. Frequently children from elementary school on will spend hours completing homework. Many schools do not structure the school day so that kids can complete these assignments in class.

Adults would never put up with this. They would go on strike. Karl Marx and Henry Ford conspired to shorten the working day to eight hours; they were surprising bedfellows. But kids are not protected by unions, nor do they have many civil rights. Homework has been expanded to fill the time available. We tend to believe that kids will waste their free time, with video games, television, drugs, or just aimless slouching and slacking. As well, we are worried that the United States has fallen behind Asian and European countries in educational attainment, particularly in math and science. Homework would appear to add intellectual value for most American kids.

Sleep Deprivation

Everyone knows that most people, kids and adults alike, do not get enough sleep. The deficit of sleep stands beside the so-called obesity epidemic as a major public health concern. Kids remedy this with Red Bull and other sources of caffeine, and they eat diets full of sugar and fat, which temporarily stimulate them. Nicotine is also used as a stimulant. Kids are deprived of sleep because they stay up too late doing homework and other structured activities, such as soccer or music practice, and because, once these chores are completed, the late night hours are the only hours available to kids for freedom, and freedom from adults.

Our local high school day begins at 7:35 a.m. No teenager is able to focus or study at that hour. The elementary school starts after 8 a.m.; little kids do better early than do teenagers. The high school dismisses early, perhaps to let kids attend sports practices and work at part-time jobs. There has been talk of a "zero period" beginning at 7 a.m. I have yet to meet a parent or student who thinks that these early start times are consistent with adolescents' circadian rhythms. This is not even to consider the earliness of the hour for teachers and administrators, who, like kids, probably did not get enough sleep the night before.

The Production of ADD and Its Electronic Solutions

Attention-deficit issues are the psychic scourge of the early twenty-first century. Along with childhood obesity, ADD and ADHD receive much press notice. Legions of kids are medicated with powerful drugs such as Ritalin to get them through the school day. But some of this may be the normal kid's response to overload—of homework, expectations, stimuli, lack of sleep. My mind wanders when I am tired. Most kids go to school tired, and they cannot focus and stay on task. A better solution would be getting more sleep and starting school later. It would also help if the after-school homework load was reduced, with more efficient use of the school day allowing necessary work to be completed before kids go home. Finally, vigorous exercise helps kids and adults with attentional issues. Sitting wedged into a confining desk for most of the school day virtually guarantees that kids won't be able to focus. Giving them the equivalent of speed to help them focus only ensures that they will stay up too late. ADD kids are often energized at night and wish to sleep late. Instead, we awaken them early and stuff them full of stimulants. One wonders whether attentional issues are simply responsive to the crushing pressures on kids, who cannot keep their minds from wandering. My English teacher in high school once embarrassed me (a probably undiagnosed ADD person) by pointing out to the class that I was staring at the clock during her last-period class. She was right. I was willing the minute hand to move forward so her god-awful class would end and I would be released from the orthopedic confinement of the school desk. After class, she told me I would never amount to anything and that I would fail the English AP test. Later in the summer, I learned that I received a 5 (the highest score), but I disciplined myself not to let her know. If she is reading this, by some remote chance, she now knows the real story: I was driven to distraction by her boring class! (South Eugene High School, spring semester, 1969 . . . name is Ben Agger . . . you know who you are. By the way, thanks for motivating me to do well on the test. And thanks for inspiring me to write this book.)

The Young Self as a Resume

Kids don't have down time because parents and educators encourage them to "build their resumes" beginning even in elementary school. These resumes will get them into college, graduate school or professional school, and then their first job. Lucky parents may receive tuition relief because of their kids' efforts. The total package includes not only grades, test scores, and class rank but "activities," the ensemble of extracurricular involvements of kids, from music to sports to school newspaper. There is debate about whether the best colleges want well-rounded people, who do well in many things but don't excel in any one thing, or whether they want kids to be lopsided, dedicated to doing one thing with excellence. In any event, kids are always "on." Their every activity, in and outside the classroom, is carefully evaluated, scored. I am calling this "performativity"—the way adults turn every kid activity into a measurable, comparable outcome.

This sends the message that things are only worth doing if they can fit on the resume or in the college application. Another message is that finishing second or worse is not as worthy as winning. Perhaps the deepest message is that kids should be utilitarian, engaging in activities and performances only as a means to an end. Parents, worried about our declining and competitive economy, want every childhood waking moment to be spent in productive, gradable activities. It is only worth doing if it goes on the resume.

The School Day

Early In

I have talked about how school starts early, too early. The day is then carved into little blocks, punctuated by school bells, with scant hall time allowed for passage between classrooms. Lunch is usually hurried. The school is a factory; product is being turned out on a schedule. The schools' product is the marketable self, the member of a future adult workforce. School scheduling sends the powerful if subtle messages that punctuality is prized, idle hands are the devil's workshop, and that life is best compartmentalized into rigid temporal units or blocks.

The school, while a factory, is also a prison. Starting school early allows school authorities to deal with a docile uniformed population. The carefully regulated school day enables social control. Teachers are always supervising their young charges, even in the cafeteria. Most schools today have strict codes of discipline, with serious punishments for rule infractions. Harmless pranks can get kids sent to "turning point" schools or even expelled. Some violators are criminally charged. My son found a little pen with a pointer device in his before-school orchestra practice, which was supervised by a parent. With no teacher around, he pocketed the pen so that he could turn it in to his orchestra teacher later in the day. She was one of his favorites. During the next period, in a science class taught by a rigid authoritarian, he was showing another boy his loot. The teacher freaked and accused our son of using a "laser pointer" to harm the other boy's eyes. We received notice from the school. Our son was exonerated when it was found that it was not a laser pointer but a harmless little device. When he returned home he asked me if he did anything wrong. I said "of course not." You did the right thing; you were turning it in to the orchestra teacher. The adult authoritarian teaching the science class sent us an email chastising our son for his dangerous and disobedient behavior. He was earning a 99 in her class at the time. The damage has not been lasting; he is majoring in physics in college, but perhaps because one of his (few) favorite high school teachers was an inspiring physics teacher. He is available to students, dresses down, and reads the *New York Times* at Starbucks.

Prison Bells and Punctuality

When the Industrial Revolution began, there were no whistles or bells indicating the start and end of the working day. As they did during the Middles Ages, people allowed themselves to be regulated by dawn and dusk. The starting bell began the process of rendering work abstract: Workers were not valued for the particular work they did; their work time was averaged as their unique skills were transformed into anonymous "labor power." Similarly, schools are factories, and the output is, directly, school work and homework and, indirectly, the industrious selves of youngsters. But, unlike some industrial workers, their time is not measured by a single starting and stopping; they hear many bells during the day, regulating their movement and their repose in classrooms. Teachers are not trusted to know when the lesson is done; they, too, are workers and prisoners of a sort. One of the problems with this factory-like regulation of time is that the assembly-line continues next hour and next day, but you cannot just stop an algebra lesson and then easily pick it up later.

Assembly-Line Education

And so, in a sense, schools are assembly lines; they are more factory-like than different from factories. Worker/students' work is specialized, subdivided, regulated. Time and motion are regimented, with bells sounded and school desks arrayed in classrooms. The aim is the regulation of labor and the enhancement of labor's productivity. It has been decided that the products of schooling, by performative standards, are grades and test scores. I notice that the product of children's labor, utterly uncompensated, is homework and school work—and ultimately of their own productivist selves—but the official view of school output is quite different. Like factories or businesses, schools are to be evaluated and compared according to students' performance on standardized state and national tests. This is also a way of evaluating teachers. Missing is the evaluation of how the self—a vague concept, for educational administrators—has been improved, strengthened, grounded, and rendered more worldly and compassionate.

Idle Hands

Like factories since the early-twentieth-century era of "scientific management" and Taylor's time-and-motion studies, schools allow kids and teachers little downtime. This would seem to maximize productivity, but in fact there is much idleness during the school day, especially inefficiencies in classroom teaching and learning. Kids could in fact accomplish much in the way of homework if the school day was more efficiently organized. But schools as factories care less about freeing kids' time after school than about ensuring that students arrive at and depart from classrooms on schedule. The goal here is social control, denying students and teachers opportunities for open-ended spontaneous learning. Teachers, for their part, must cover certain material so that students can pass the standardized tests mentioned above. Everybody seems busy, but students' stories about time use during their school days suggest that things could be organized quite differently, allowing more free time after school.

Broken Backs and Contorted Bodies

Desks

Foucault suggested that learning cursive writing is an exercise in discipline. The child must master certain fine motor skills that, today, are nearly useless in the age of word processing. Schooling involves both bodies and minds, much as prisons and factories exact a physical toll. Notable about schooling is that the classroom, the primary site of instruction, does not encourage standing up or roaming but requires students to sit in chairs at cramped and uncomfortable desks. Although kids walk through the halls between periods, and may exercise during P.E., for the vast majority of the school day kids are not only inactive but cramped. Desks are a contemporary version of medieval instruments of torture. They ensure that kids will end the school day tired, lethargic, possibly out of skeletal alignment. The teacher gets to walk around. Students are punished if they leave their desks. Adults have long since repressed memories of the confining desk. They should sit in one for an hour and then multiply that experience by seven or eight to get a clear sense of how schooling, on the prison/factory model, takes a physical toll.

Backpacks

My daughter's high school backpack weighed over twenty pounds. She also lugged around a viola. My son carried a similar backpack, pulled a full-size cello, and shouldered a large, loaded tennis bag. These loads are debilitating for the musculoskeletal system, and we are talking about kids who do not possess adult strength and have finished growth plates. My kids carried heavy loads even during elementary school, where they needed many textbooks as they rotated from subject to subject in different classrooms. They were not allowed to use rolling backpacks or bags of the kind that flight-attendants use. This was against school policy. The use of heavy backpacks and other loads does not promote physical well-being and is not a suitable form of "resistance training" (such as weight lifting). It is like asking factory workers to carry their own tools and machine parts.

Handwriting

As I noted, Foucault examined penmanship as a crucible of fine motor skills and thus of discipline. Imagine sitting still in a contorting desk and then having also to produce "neat" writing, whether printing or in cursive. Grades are taken off for messiness, ignoring the fact that adults are guaranteed a "neat" outcome with word processing and printing. Some kids are wired not to be able to write neatly (I am in that group). For others, perhaps artistic types, neat printing and flowing cursive come easily. These are tests of character and of academic suitability not only in the early grades but later in junior high and high school. I always told my kids that most successful people I know have very sloppy handwriting; they are too busy to care, and they can readily retype their work. As well, kids with learning disabilities simply cannot earn good penmanship grades. Interestingly, handwriting is evaluated both in itself and in all sorts of assignments in different substantive subjects. This is a minefield for the hurrying, creative, possibly ADD student. And, now with word processing, it has no carryover into the adult world; it is not a skill that needs much attention.

Decorum/Discipline

Imagine an active boy (or girl). He is what is called "motoric"—he likes and needs to move frequently. He is wedged into the confining school desk and expected to sit there for years on end. Youthful bodies rebel; they fidget restlessly. Sometimes kids need to stand up and stretch. They need to roam. All of these behaviors are sanctioned by teachers and administrators who feel that they cannot afford to have a school full of wanderers. Longer exercise sessions are rejected because school authorities feel that they must cram all they can of academic classes into the school day and also because exercise is supposedly compartmentalized in P.E. and team athletics.

Bodies in motion learn better. The mind-body connection is fundamental. Healthy brains depend on exercising bodies. Endorphins are produced in exercise and there are many other biochemical sources of bliss and well-being triggered by movement and aerobic stimulus (Bergland 2007). Teachers who exercise know this, but nevertheless kids are still strapped to their desks. Movement is punished as lack of decorum, even as disobedience. I am convinced that most kids who fidget and move do not intend to be "bad"; they are just answering to their bodies' need for movement during a long and tedious school day. Boredom causes ennui, as does confinement by those wretched desks and chairs.

Bodies in motion burn calories and achieve cardiac fitness. We cannot complain about the flabbiness and lack of fitness of American youth and at the same time confine them in medieval torture school desks—and punish them for transgressing. Well, we can complain and confine simultaneously. That is what we are doing today. The mania about test scores and losing America's comparative advantage trumps public health concerns.

Just the Facts: Minutiae and Memory

The Third Longest River in Brazil

When we helped our kids with their elementary homework, we usually found ourselves reading lists of vocabulary words or geographic trivia—the third longest river in Brazil, the second staple crop of Idaho (first, probably potatoes), the Texas general at the Battle of San Jacinto. The fact-based philosophy of education of George W. Bush and the wider conservative movement rejects theory, multiculturalism, international studies, and the teaching of evolution. We parents did many of the same things; I remember memorizing as a staple activity of studying. And this is not only a memory project. The fetishism of unconnected facts invades not only testing but term paper writing and larger projects. Perspective and point of view are rejected in favor of what is purported to be objectivity—the many minutiae that must be mastered in order to make the grade.

Small Picture versus Big Picture: Aversion to Theory

I read the fetish of facts as opposition to theory, big-picture thinking. Thomas Kuhn (2012) argued that knowledge really proceeds not from fact-gathering but from "paradigmatic" shifts—shifts in the large models we use to understand the physical and social worlds. A few facts here and there did not allow Copernicus to overtake Ptolemy or Einstein to best Newton. Large-scale theorizing, painting a sweeping picture of various aspects of physical reality, achieved progress in science.

Paradigms are dangerous, and Kuhn revolutionary, because they imply relativism, which threatens the master narratives of religions, positivism, and capitalism. That there can be different viewpoints on fundamental issues such as the origin of the universe challenges fundamentalism—the idea that there is a single truth and it is disclosed only to the faithful.

Curricula and the Culture Wars

The culture wars, as they are called, stem from the 1960s, when young people began to demand relevance in their higher education and minorities developed knowledge, including about themselves, based in "standpoint." The 1960s gave us partisan knowledge, which originally stems from Marx and Engels's Eleventh Thesis on Feuerbach—that the point of knowledge is not only to understand the world but to change it. The culture wars began in the 1980s, during Reagan's presidency, when the Right organized itself to reverse certain important gains of not only the New Deal but the 1960s. Attacks on civil rights legislation, *Roe v. Wade*, busing, and affirmative action have mounted since the Reagan years. In the twenty-first century, the pendulum has begun to swing in the opposite direction in important ways.

These issues are being fought in Bible Belt school systems as school administrators and politicians seek to ban the teaching of evolution in favor of "intelligent design." Writ large, the culture wars take the form of a politics of curriculum— what will be taught, which textbooks will be used, how testing will be conducted. Another aspect of these culture wars as they affect public schools is the issue of school prayer, which is widely supported by the Right. These are not settled issues.

The Performance Principle I: Compulsory Evaluation

The Tested and Graded Self

I remember grades from my public school days over forty years ago. But our kids are graded far more compulsively than we were. Nearly every school day brings gradable activity in a host of their classes, all the way from testing to the evaluation of in-class assignments and homework. Teachers keep computerized track of student grades, and they use these programs to calculate final grades. This falls under the overall project of bringing human activities under the rule of evaluation. Occasionally, our kids will receive a "completion" grade, for just turning in the assignment or project. Parents and kids love completion grades because they lower the stress and raise the child's average.

Kids internalize all of this and view themselves in terms of their report cards. Daily, their identity depends on the myriad gradable performances, from academic work, to art and music, to P.E., even to lunchroom comportment. I remember our kids coming home from elementary school with a "behavioral" grade; a check mark in their daily planner indicated that they had "misbehaved" that day. A double-check was a no good, very bad day! That was usually accompanied by a teacher's rebuking comment. Our kids would often tell us how they were "written up" for violations that either seemed incredibly trivial or were downright unjustified—like the day when my son, fearing that his whole class would "get in trouble," admonished a classmate for goofing around in the hall. My son was apprehended and written up, losing recess. I urged him to talk to the teacher the next day and tell him what really happened. To his credit, he did, but the teacher told him that he couldn't change the check mark. I didn't tell my son that the teacher probably didn't believe his explanation. Every parent in America knows about these minor but continuing injustices that represent the "adultist" domination of children, who have few civil rights.

The Ubiquity of Evaluation

Confronting daily, weekly, monthly, six-week, and semester-long evaluation puts self-esteem in play. It is also highly stressful for academically oriented kids or even for marginal students (perhaps more for marginal students). For these kids, school is a series of traps. They lug around heavy backpacks from class to class; they must keep their assignments, homework, and tests straight; they must remember deadlines. All of this adult-like multitasking begins in elementary school and only gets worse as the high school years arrive.

Not only is this stressful and threatening to identity and esteem, It is also disempowering; teachers have all the power and kids have none of it. Most high school kids I know mistrust teachers and all adult authority; they realize who holds all the cards, who is doing the endless evaluating and disciplining. As all parents know, kids frequently bring home low grades on particular gradable units with absolutely no explanation attached or included. The child must then muster the courage to visit the teacher after or before school on the next day. Occasionally, parents get involved to contest a grade and, as most of us know, that usually does not end well for the child. Because we are college professors, we try to stay out of the fray. But sometimes the grading has been so blatantly unfair and even incorrect that we step in. The teacher almost always wins because it is a case of "he said, she said."

Progress Reports

In our school system, these reports are delivered every six weeks. Low grades result in temporary exclusion from orchestra, band, and team sports. But our kids received three-week progress reports, as well. Sometimes, these mini reports are misleading because grades are missing or few gradable units were included. Every few weeks, then, in addition to their mountainous daily evaluations in all subjects, children can expect to receive a global evaluation of their performance and comportment. In our school system, these reports are preceded by an ominous recorded telephone message left on our machine that says, in effect, the grades are coming; don't let your kid hide them from you. Everyone's stress rises (including that of teachers, who can expect complaints from pushy parents).

As I said, we have occasionally intervened on our kids' behalf. We have largely given that up because it is so de-energizing and leaves us cynical. We wonder how schools of education can be turning out so many authoritarian anti-intellectuals. We worry that our kids will end up hating school and learning generally. I asked our daughter which teachers she really liked and respected in junior high and she thought for a long time and came up with a few names—and she had several fingers left on one hand. I asked our son during his first year of high school whom he felt he could approach with a problem and who would respect him. He responded "the tennis coaches" (who are also teachers). I and my wife are educators, too, and we feel solidarity with K–12 teachers, who have little academic freedom and intellectual room to move. School teaching is usually alienated labor. Many teachers do their best and identify strongly with their students, but many others deal with their alienation by blaming their charges and dialing up the discipline. If authoritarianism abounds in schools, it needs to be traced to authoritarianism in society as a whole, which opens up the perennial issues of rulers and ruled, capital and labor. Blame is beside the point; we need to change large structures, but in ways that do not bypass the small details of what Foucault called the microphysics of power, including the learning of penmanship, desks that confine, dress codes, rigid curricula.

Asymmetries of Power

The basic problem of the evaluation of kids, including college students, is that they have no power, no recourse, no options. Once the grade is assigned, it haunts the child forever, becoming part of their permanent record. In our credential society, parents overemphasize the importance of getting into certain colleges and thus they redouble the pressure already experienced by young selves who are caught continually in the cross-hairs of evaluation. It is easy to fall into this trap: Parents know that, on some level, grades matter. Kids have been told this from day one. Teachers and coaches reinforce this. Evaluation occurs almost daily, and then at three weeks and then at six weeks and then at semester's end. There are so many minefields; kids can go wrong anywhere and receive low grades, even if they have only misplaced homework or been late on an assignment or two. Our own experience is that just being sick for a couple of days exposes our kids to academic backsliding; they miss class work, homework, tests. Our daughter had to make up a ninth-grade English test for missing class because of a school-sponsored tennis tournament. Incredibly, the teacher put her out in the hall to take the test.

It is difficult to imagine adults putting up with these indignities, including public rebukes from teachers who express aloud that certain students are slackers and failures. We would change jobs, try to get our boss fired, start drinking heavily! We would join solidarity with our colleagues and perhaps start a union. We would work to rule, slowing down our pace. We would take frequent sick days off. We might change careers and return to school. Adults' workplaces, in our post-Fordist era, have become somewhat "flat" in their authority structure; they are warm and fuzzy; collegiality and a first-name friendliness are stressed; we can dress down on Fridays and perhaps on other days, too. Children's workplaces, by contrast, operate on a model of authority and discipline taken from the eighteenth and nineteenth centuries, when workers had no rights. As I said above, children are the last neglected minority group.

The Young Self as a Resume Revisited

Compulsory evaluation is defended as the price of college and career preparation. Kids are told that it is never too early to begin assembling the college resume, with good grades, diverse activities, and a network of adult sponsors, including teachers and community members. In a sense, this is true. Admission to elite colleges and universities is highly competitive, and there are many applicants with 1400-plus SAT scores and nearly 4.0 GPAs. Parents looking for an edge overplan their children's lives and have them enrolled in all AP classes and then in after-school and evening activities designed to attract the attention of college admission officers. The young self becomes her resume, just as her parents have been reduced to the same status in their careers. The difference is that when I went to college (starting in 1969), this self-assemblage did not begin to take place before high school, and even the latter part of high school. Today, it begins in first grade or even before.

The Performance Principle II: Testing

Standardized Testing

Standardized tests may include the SAT and ACT and also various mandated state tests that purport to measure school performance. Underperforming school districts may be punished economically. The point of standardization is to allow for comparison, especially where there may be significant differences in context. A school with a lax grading policy may underperform on standardized tests compared to a more rigorous school with lower overall grades but higher test scores. Similarly, standardized tests used for college admission and placement allow one to control for differences in race, gender, family income, region, and so on.

The problem with all this is that sociologists generally believe that these standardized tests are themselves highly flawed and lack genuine objectivity. They are laden with class bias, rewarding kids who attend private schools and well-funded public schools. So-called intelligence tests have the same built-in biases. As well, SAT scores predict little more than first-year academic performance in college. After the first year, the test results lose significance. These tests are a proxy for social class background and the academic milieu of one's high school. Using them only reproduces the elite who dominate the admission list of elite colleges and universities.

The Tested Self

What, exactly, is being tested in the compulsive routines of school evaluation? On one level, the student is being tested on how much he or she knows or remembers. On another level, the self "itself" is being tested, measured, evaluated for worthiness. I have commented already that self-esteem hangs in the balance. There is something peculiarly punishing and bruising about being constantly evaluated. Young selves learn that the activities that "count" are those that count toward one's GPA, test-taking ability, or college admission. Utilitarianism rules the school. Extraneous knowledge, knowledge "not on the test," is not worth memorizing or learning. The young person quickly realizes that he or she will be put to the test every day of their youthful existence—in the classroom, on the sports field, in music competitions. Second-best is not good enough. The tested self fails more often than she wins, necessarily. This is a formative experience for later life. Being constantly tested provides a value framework, a worldview.

The Pre-Labor Force

Parents may rarely think of their kids as employed, except in part-time work after school or on weekends. But kids constitute a work force. They do regular hours, many more than forty per week if homework is included. They are regimented and supervised. Their time is closely monitored. The produce "output"—tests, assignments, homework, projects. But kids' work is unpaid, and it occupies more hours of the day, including nighttime. Kids constitute a pre-labor force that learns the experiences of working, duty, discipline, and fatigue. This prepares them for adulthood, it is thought, and it diverts them from idleness. As well, it keeps kids occupied when parents work long hours. There is a babysitting function in play here.

Few adults would volunteer to perform unpaid labor, except, at the margin, some volunteer work such as PTA involvement. But kids have no choice; they are compelled legally to attend school and perform the various tasks assigned. Schools have three layers of curriculum, moving from superficial to deep:

- What is learned from the textbooks and lessons, in math, science, history. This is the *explicit curriculum.*
- What is learned about adult habits of punctuality, respect for others, team play, hard work. This is what Bowles and Gintis (1976) call the "hidden curriculum."
- What is learned about how life is a grind and play and freedom rare; how people are judged by their performances, which are evaluated and compared. These are *lessons in deferred gratification, endless productivity, and performativity.*

Teachers and school administrators are well aware of the first two levels. The third level, performance, is concealed from adults, but children know it well. This is the level on which children become a pre-labor force. They are not just preparing for paid work, as they are on the second level, but already working. Kids are no longer kids, if you examine the backbreaking toil of carrying backpacks and completing homework. Somehow, we have eliminated childhood.

Bodies in Motion: From Play to Sports

From Recess to Team Work

Everyone knows that American kids do not get enough exercise, an issue I pursue in Chapter 12. They also eat unwisely, preparing for adult illness. Unfortunately, much of this can be traced to the crowding of kids' days and nights. They do not have much time for exercise or self-directed play. We used to have recess, where kids would devise their own games and run around. We have substituted for that team sports, which are taken very seriously by adults. Indeed, some adults think that their athletically gifted children will earn college scholarships in their sports.

Team sports are valued both because professional sports (and proto-professional college sports) are important in this society of the spectacle (Debord 1970) and because teamwork is valued as a workplace virtue. "Going out" for school sports is an essential feature of resume building for young selves. Sports, play, and exercise are not the same. Kids in team sports can stand or sit on the sideline and even practices have much down time. Play, by definition, is self-directed and outside the purview of coaches and other adults. Exercise, which can be derived from team sports and from play, is vitally necessary for young minds and bodies, and it is best experienced as play, having no intrinsic purpose. The value placed on winning (or, perhaps better, avoiding losing) is inimical to play, which is its own reward. Schools do not want kids to have recess because this would give them dangerous freedom, just as too much time spent in the prison yard could have untoward consequences.

Organized sports, especially for boys but increasingly also for girls, are marks of character, achievements in their own right. Kids are stratified into leagues of varying ability levels. Winners and losers are determined. And individual statistics are kept. Kids declare for one team or another in much the same way they go out for the school newspaper or orchestra. These are notches on the belt, lines on the college-bound resume. Unlike when I was a kid, children's sports today have heavy parental involvement, the better to bend sports to the rule of performance. Every kid who does sports understands the sting of the coach's rebuke, his public humiliation of kids who fail to measure up. Not only are resumes built but self-esteem is diminished except for the lucky few who excel athletically. And even they will taste the bitterness of defeat frequently. Almost everyone loses in organized sports.

ADD as the Absence of Aerobic Activity

Aerobic activity, stressing the cardiovascular system by raising the heart rate and requiring the person to process a lot of oxygen, is linked to the resolution of attentional issues and mood disorders (Ratey and Hagerman 2008). Brain chemistry is affected by how much or how little the person works out, especially in endurance sports. Although medication may also be needed to address attentional and mood issues, it is now widely recognized in the medical community that physical activity reduces the need for drug therapies. It is no accident that the epidemic of ADD and ADHD diagnoses overlaps the reduction of time that kids, both in and out of school, spend in vigorous physical activity. They are indolent, crammed into desks; they are sleep deprived, and thus stimulate themselves artificially; they are stressed and eat high-sugar and high-fat food and compound their ill health; they drive and don't walk; they simply don't have time for play and exercise. The obesity epidemic will never be checked without reducing homework, the overscheduling of kids, their premature resume building, and their overall participation in the pre-labor force. As well, kids must have adult role models, both among parents and teachers.

Parental Pride and the Performance Principle

Every kid knows that the trophies they win in sports or other competitions are for the parents. The parents get bragging rights, as the many out-of-school private lessons and select league participation pay off. Parents lard their own resumes by producing kids who win; coaches acquire esteem through championships. Although kids care, they care more about just having fun and good fellowship. And parents who care too much produce stress for kids, who then experience their sports as yet another obligation, yet another opportunity for possible failure.

Fast-Food Cafeterias

School Lunches and the Standard American Diet (SAD)

Just as everyone knows that kids have too much homework (Kralovec and Buell 2001), are overscheduled, don't get enough sleep, and don't exercise enough, so everyone knows that school cafeterias are not part of the solution; they are part of the problem (Schlosser 2001). They offer food high in fat and sugar, and some schools allow fast-food franchises in their buildings. And if kids go off campus, they patronize fast-food restaurants where they replicate the diets of their parents and what they see on television and in the media. There is little attention to healthy diets, just as there is inadequate time devoted to real exercise and vigorous play in schools. High school kids subsist on French fries, pizza, burgers, Cokes, and Red Bull. They get little roughage, few complex carbohydrates, and inadequate lean protein especially derived from vegetable sources; they are overstimulated by sugars and caffeine. Our younger generations are floating on a sea of high-fructose corn syrup, purveyed by corn growers, supermarket chains, fast-food franchises and the frozen food industry (Pollan 2006). Healthier food just doesn't taste good to the average teenage consumer.

Stimulants and the School Day

Like adults, kids caffeinate and stimulate themselves in order to get up in the morning and then get through the long school day. Cokes, Red Bull, and Monster are the beverages of choice. These have empty calories and too much caffeine, which, over time, can lead to teenage hypertension. Part of the problem is sleep deprivation. Another part is the early start time of schools. And kids stay up too late both to complete assignments and to unwind.

Fast Food and the Adolescent Driver

Perhaps less than a third of my high school class had wheels. Today, most kids drive to school once they are old enough to have licenses. They then drive off campus for lunch, and after school they indulge their appetites for fast food, sugar, stimulation. They mimic their parents, who also lead driving, shopping, eating lives. Adolescent driving is a serious problem from a number of points of view, including where and what kids eat. It is dangerous when mixed with alcohol and drugs. But in a deeper sense a nation of sixteen-year-old drivers is a nation in which childhood has been attenuated and a developmental stage skipped.

Many a parent has remarked that they cannot wait for their kids to drive so that they will be unburdened of chauffeur duties. This promotes a lack of accountability as well as deep-seated feelings of abandonment. Some parents purchase late-model cars for their kids, which become a status symbol for both parent and child.

Discipline and Punishment

The Penal Colony

Every school child and most parents know that misbehavior in school has consequences, all the way from a bad report card to after-school suspension to missing recess. At our kids' former elementary school, kids who misbehaved during the school day were made to stand by the fence watching the other kids have recess. This public humiliation reminds one of the stocks in Puritan New England. It also denies motoric kids precisely what their writhing, restless bodies need: play and exercise. I often wanted to put the teacher on the fence, too, for assigning this punishment. Frequently, as I drove up to the carpool line, I would notice not one or two kids on the fence but legions of them. Clearly, collective misbehavior was being punished.

Children in fear lose confidence in themselves, resent adults, and dislike school. Most kids at one time or another profess that they hate school, and they acknowledge that school is prison-like. They are inmates and teachers wardens. School is experienced as a life sentence. Adults' authority in most schools is nearly unchecked. Teachers and administrators have broad latitude in their abilities to identify and punish wrongdoing. Kids walk on eggshells, worried that they are being too noisy in the hall, their shirts aren't tucked in, they are boisterous at the lunch table, they laugh too hard at a joke told during class time. Since parents aren't present, adult authority in the school setting is frequently arbitrary and experienced as arbitrary by kids, who have no court of appeal. Although parents do get involved in disputing the facts of their kids' cases and questioning the severity of their punishment, this is always after the fact.

My daughter, then in junior high, once called me to say she was found to be out of compliance with the dress code by a teacher who was not her own. She worked as an aide in the principal's office and the last time she wore this particular outfit (conservative by any standard), the principal approved it. The issue was weighty: whether she was wearing cargo pants (prohibited) or not. She was pulled out of class (math, as I recall) and made to change clothes on the spot; she was given some prison-issue-like garb. She called me at lunch time and I went to the school, visited the principal, and wrote a brief letter requesting that she be removed from the strictures of the dress code. He quickly agreed. I promised that she would continue to dress conservatively, but without the corporate (or prison-like, depending on perspective) "spirit shirts" that were required. My daughter walked with a lighter step the rest of the school year. She also got to wear bright pink, then her favorite color.

Race and Class

As I mentioned earlier, our school district just resolved, in a close vote, to table the proposal to move toward standardized dress in high schools. This was a veiled move to ban baggy pants pulled down below the butt. Standardized dress is a plan conceived by people who somehow believe that boys won't also pull down their uniform khakis! Uniforms may blur disparities in wealth between students' families, and some schools use uniforms as a means of reducing bullying. But simmering underneath these justifications are the issues of race and class. Most high school kids sense that there is a racial aspect to dress codes, that they're partly about making black kids white and also preventing gang-like rebellion against authority. I have no brief for such pulled-down attire, but the connection with race and also with social class seems obvious. Schools want kids to look preppie, attired in Abercrombie. Preppie attire sends two related messages: It is not gangsta attire, and it is the chosen uniform of kids learning how to dress for success.

Adolescent kids not encumbered by a school uniform or standardized dress policy experiment with style as they experiment with their own identities. They make mistakes, by adult standards, but this is part of growing up. My dad hated my hippie-movement clothing during the 1960s, and I eventually turned out all right! During the late 1960s, when I attended high school, the burning issue was long hair on boys. Today, boys are allowed to wear their hair long (that could be preppie), but body politics involve issues such as where on the butt the belt-line falls, whether the jeans (probably distressed) have too many holes, and whether the girls' shirts are too revealing.

We Are Not Uniform

Adults wish that kids were packaged similarly; this would make them lower maintenance. But kids, especially teenagers, are in search of identity, and, as they do with clothes off the rack, they must try on different versions before they settle on the right one. This may take many years beyond high school. This process doesn't halt with college graduation; adults, especially those who change careers and divorce, also reinvent themselves. We should encourage this among our youngsters and not use a factory standard of product uniformity with which to clothe and treat the children.

Making Reading and Writing Routine

Grammar and Authoritarianism

Linguists have mistrusted structural grammar for decades. Yet we continue to expose students to gerunds, participles, and diagramming of sentences. It is helpful to know the basic parts of speech and writing—noun, verb, adverb, adjective. But structural grammar not only does not teach how to write or read; it kills the spirit and deadens kids to literature. Why are grammarians so out of date? I suspect that the esoteric code of grammar allows English teachers to keep control of young minds, who otherwise might explore the delights of Twain or Toni Morrison. They might write their own stories and scribble outside the box. They might find their muse and learn to be free—two key lessons that education can provide.

How to learn to write, then? By imitation, by reading a lot and then attempting one's own stories. I learned that way, and you probably did, too. I remember that my seventh-grade grammar teacher was cute, a redhead, but I cannot remember a single thing I learned. Probably not her fault. I was dealing with early adolescence, including having a crush on a girl with white flats. And I remember watching one of the early NASA space shots on her classroom television. I do remember something about her class and its fixation on the diagramming of sentences. I never got it right the first time and so my final sheet was smudgy with erasings, perhaps an early lesson in the undecidability of the text—the way that every writing could have been done differently.

Gestalt Readings or Detailed Deconstruction

Even in my kids' honors English classes, they were tested on the minutiae of stories—who did what to whom in Chapter 7. Instead of commenting, in gestalt fashion, on the whole story, they are forced to deconstruct it into its details. In elementary school, our kids participated in an accelerated readers program, in which they would read kids' and young adult books and then sit at a computer and take tests on the details of plot and character. There were ten questions, and their scores would be totaled over the course of the year. My son, in fourth grade, as I recall, won the award for the whole school—an occasion of bursting parental pride! (I say this with tongue in cheek, totally.) This was of little use to him. What was useful was simply reading dozens of books whose style informed his own development as a writer. As a teenager, he writes more elegantly and creatively than the vast majority of my college students. No teacher should take credit, and nor should I or my wife. He learned by imitating. That's how it works.

Does a Writing Tree That Falls in the Forest Make a Sound?

Writing trees—schematic diagrams of how essays should be organized—may be useful as a rough guide. But this hems in kids, although it allows teachers a template with which to grade the work submitted. Too many writing teachers—who don't themselves write—don't understand how writing works, through a combination of imitation, daydreaming, serendipity. I'm not blaming them; virtually all teachers are underpaid and overburdened. Pay school teachers as much as college professors and require at least a master's degree. That will overnight upgrade American public education. And I'm not also implying that most professors write. I already mentioned that the vast majority of college professors don't write for publication, either.

"I'll Never Read Another Book"

One of the great paradoxes of the literary instinct—the sheer urgency to write and reveal—is that schooling often kills the desire to read and take literature seriously. And yet, as I have been arguing throughout this book, kids write all the time—texting, posting, messaging, tweeting, blogging. We haven't yet squelched the literary impulse, although grammar has come close. It is fashionable to blame everything in the way of cultural decay on youngsters too lazy to get off the couch, where they play video games in casual repose, and get to the library, where they can sit in a straight-backed chair and read a book with dense text and no pictures. But we who teach writing are largely responsible for this rechanneling of kids' literary energy into unofficial venues and forms that we allow to count as literature.

Sometimes we have to live with paradoxes, as we must here. Kids hate school, but love writing, yet not the sort valued in schools. They have the protean impulses to express themselves and to connect, for which social networking and blogs are perfect. Adults have these impulses, too, which is why we also write, and sometimes furiously. We join message boards, blog, email, and text. Self-publishing has exploded in recent years, providing writers with new opportunities for reaching an audience. Words are spewing forth, which is both good in itself and potentially problematic—if the words aren't informed by reading, cosmopolitanism, progressive values, a sense of style and grace. You can write your way out of Peoria, and perhaps staying in Peoria is better than life in Manhattan.

Much of the writing going on—think of the billions of words and images clogging the Internet, let alone libraries—is born of joy, anger, passion, play. It is writing done for its own sake, not particularly worried about finding an audience. Even then, it is difficult to stay invisible on the Internet. People with time on their hands will find you and join the conversation. And the Internet affords global reach; transaction costs are very low. It is flexible and permits multitasking, perhaps making the lonely writer's garret a thing of the past. Writers aren't lonely because some have fan pages, linking them and readers. And many are on Facebook. Literary life increasingly resembles a latter-day Star Trek convention, with creators and idolators meeting regularly along the information superhighway. It is difficult to hide in plain sight as publishers, studios, and record labels take to the Internet, and, in any case, the culture industry is increasingly a celebrity culture. Even a few years ago I wouldn't have imagined that Kim Kardashian and Barack Obama would both use the portal of Twitter.

Sputnik, Globalization, and Math Mania

From Newton to Sputnik and Beyond: Why Math Doesn't Matter

The Enlightenment, which began early in the 1600s (with roots in the earlier Renaissance), allowed scholars to know the world directly, not only through the mediation of various good books such as the New Testament, Koran, Talmud. This was called *empirical knowledge*, based on sense perception. Europe and England advanced quickly toward the Industrial Revolution a hundred years later once people were freed from blind faith. If you are sick, do you visit a doctor or pray? For the Enlightenment, it is a no-brainer.

The most "positive," perfect, unassailable form of knowledge for the Enlighteners such as Newton was mathematics. Math is precise, is objective, and allows for statements about causality, especially with respect to nature. Einstein muddied this when, in 1905, he said that math and science aren't precise but depend on the place and time of the observer; he introduced the concept of relativity. Newton supposed that one could stand outside the world and be perfectly objective and causal about nature, especially when one sought Truth along the royal road of mathematics.

The Soviet launching of Sputnik in 1957 led American leaders and educators to redouble our emphasis on math and science in schooling, but not for Newton's philosophical reasons. We wanted to stay ahead of the Soviets, especially in the race for space. We were falling behind after World War II. We also wanted to stay ahead of them in nuclear weapons, a doomsday scenario. Newton and Sputnik, taken together, have led American schools to emphasize math perhaps above all else. This math mania leads educational testers to give equal weight, for example on the SAT, to math, analytical reasoning, and now writing. One cannot get into an elite graduate school without doing well on the math part of the GRE, which requires one to bone up on high school algebra and geometry.

Math Theory versus Getting All the Steps Right

In our kids' junior high, kids were not allowed to use calculators, which are permissible in most college math classes. Not only is math important at this school, as at most others. It was to be done in an old-fashioned way, maximizing children's pain. And part of this old-fashioned way was to stress that every single detail of the various steps of the math problem must be done correctly. A single error would deny any credit for the correct answer. Conceptual understanding of math—in effect, math theory—was to take a backseat to the anal compulsions of rule-bound math teachers. If kids didn't hate math before seventh grade, when they entered this junior high, they would quickly learn to do so. And, again, the scolding teacher would become a source of resentment and anxiety. This only gets worse in high school.

Do Jobs in the Twenty-First Century Require High School Algebra?

Every kid suspects that they won't need all of this math later on. Most jobs don't require it, although it is useful to know some math theory, about algebra, geometry, even inferential statistics. Since Newton and especially since Sputnik, the learning of math has symbolic value, standing for rigor, intelligence, even patriotism. It is used as a sorting device, as with SAT and ACT tests. There is the implication that smart people are good at math and less-well-endowed people aren't. I have already addressed the perils of standardized testing that actually test little more than social class and family background and not innate intelligence. Math is seen to be closer to an actual raw intelligence test because it is supposedly less culturally laden than literature, history, social studies.

Learning-disabled kids (ADD comes immediately to mind) are quite challenged by math and by foreign languages. The math mania makes it difficult for these kids, who may have solid conceptual understanding of math theory but who founder on tests because they make detail mistakes in the many steps of the problems. In this sense, math is actually quite biased; it rewards those who can sit still and not get easily distracted. This begs the question, addressed above, about the multiple causes of ADD and ADHD. For some kids, thus, not getting enough exercise during the school day actually makes it more difficult for them to learn math—a surprising correlation, but one grounded in the very chemistry of the brain. We are beginning to recognize that ADD is over-diagnosed and over-medicated, playing to the needs of teachers to keep order in crowded classrooms and of pharmaceutical companies to capitalize on a growing market for psychostimulants.

Alternatives to Traditional Schooling

Parents and their kids have confronted some of these issues since the 1960s and even before. The three broad categories of alternative forms of education include "deschooling" and "unschooling," free schools, and homeschooling.

Alternatives I: Deschooling and Unschooling

Deschooling stems from the writings of Ivan Illich (1971). It involves liberating kids from school and implanting education firmly in existing social conditions, recognizing that education is a byproduct of working and living. "Unschooling" (Griffith 1998) is a related and more contemporary movement that views schools as prisons and factories. This movement questions the authority, both legal and social, of schools over children. It is implicitly a kids-rights perspective, and one that I largely share. At the root of both deschooling and unschooling is the premise that education should not be compulsory but an option. These two perspectives have clear libertarian overtones.

These perspectives don't blame inadequate educational funding for underlying problems with schooling. Although a significant social and economic investment in children's futures is important, de- and unschooling question whether a school is the proper site of education. Both perspectives view schools as essentially totalitarian institutions that stifle kids' creativity and thwart kids' spontaneous self-development. These perspectives focus on educational authoritarianism and the ways this kills the love of learning. They also, especially the Illich perspective, focus on ways in which schooling as we now know it is functional for capitalism, a veritable credential mill.

Schools of education and the teacher-certification business are especially indicted by de- and unschooling advocates. Teacher training is viewed skeptically, especially where it is assumed that education should be a separate undergraduate major somehow separate from advanced study within particular academic disciplines. The de- and unschooling crowd believe that teaching should be taught as a moral and political discipline, a way of fostering sympathy, empathy, and a sense of social justice. Teachers won't teach to the test, nor mete out discipline and punishment, nor view their work simply as providing credit hours and credentials. In Freire's (1970) terms, teaching will be a pedagogy of the oppressed.

Schools will be viewed as social and political institutions and not as neutral sites of the mere transmission of knowledge, usually reduced to facts. Perfect objectivity is impossible, and teachers should seize the day and educate children for utopia—for daydreams of the good society, to which they make a contribution. Schooling should involve volunteer work ("service learning") and involvement in such community development projects as Habitat for Humanity and the Peace Corps. Teaching will be radical in the sense of getting to the root of things, including oppression and inequality. De- and unschooling people are usually politically progressive and critique the role of schooling in reproducing both the capitalist economic order and dominant ideologies.

Alternatives II: Free Schools

Summerhill and other sixties experiments with free schools do not abandon formal schooling in the way that de- and anti- people might do. Kids attend school, but they are "free"—free to choose courses, to be themselves, to express, to create and think outside the box. Schools are communal sites of living and learning. My sister attended a path-breaking sleep-away free school in Canada called Everdale. She joined the kids of other progressive parents such as Staughton Lynd's daughter in living and learning on a farm. They did not have formal curricula or even textbooks. They moved from module to module, teacher to teacher, as they dabbled in everything. Indeed, there were no formal boundaries between teachers and students. Everyone has something to teach, on this model. And there was real egalitarianism between students and teachers, who were addressed by their first names by the students.

Montessori

Maria Montessori inspired many experiments in alternative education. Kids learn at their own pace in free-floating classrooms that do not have set curricula, texts, tests, or homework. Teachers facilitate the development of the whole child and they recognize that kids develop at different paces. Most classrooms have mixed-age groups of students. Our own kids attended Montessori, but we found that there was not enough structure for learning to take place and we also disliked the cultic belief of some teachers in Montessori as a panacea for all ills. In a word, we thought the school was overhyped and, in its way, secretly authoritarian and dogmatic. We found teachers actually quite judgmental in ways that they are not in free schools such as the former Everdale.

Waldorf

Waldorf schools, developed by Rudolf Steiner, stress children's creativity and spirituality. They are more explicitly progressive than Montessori, although they share with Montessori the idea that children develop at different rates and that the whole child must be the focus of education. Waldorf shares with Montessori the idea that kids are creative and that creativity must be fostered and not stifled. Both approaches reject standardization—of testing, curricula, teacher training. Both Montessori and Waldorf are private schools and they tend to be favored by the middle and upper-middle class who can afford them. The parents who send their kids to these schools may be affluent Volvo- or Prius-driving professionals, from the 1960s. They are grownup hippies and they cringe when they think of the authoritarianism and rigidity of traditional schools.

Alternatives III: Homeschooling

Escape from Race and Secularism

By far the most rapidly growing alternative to traditional schools is the vast homeschooling movement, which is especially prevalent in Southern Bible Belt states. Whole bookstores provide texts for use by parents in homeschooling their progeny. And now there are exercise classes, team sports, and even classes in specialized subjects at local community colleges provided for homeschooled kids. Homeschooling is not grounded in the systematic philosophies of pedagogy of the free schools, Montessori, or Waldorf. They assume that parents know best how to educate their kids, and in most homeschooling households the stay-at-home mother doubles as teacher. Typically, kids work through their guides and texts more quickly than in traditional schools, and thus kids have a lot of free time after the morning lessons end.

Homeschooling parents lament the social problems found in many urban and even suburban schools. Homeschooling is, for many, an escape from issues of race, with white parents wanting to sequester their kids away from black and Hispanic children, who are feared as actual or potential gang members. Other parents who seek racial and cultural homogeneity for their kids may move to the suburbs or send their kids to private school. Homeschooling is an option only if one parent doesn't work, which is increasingly not the norm.

Homeschoolers, especially in the South and rural areas, may be sought by evangelical and other Christians, who reject the secularism of public and secular private schools. Creation can be taught unashamedly, as can religious doctrine.

Progressive Homeschooling

Much homeschooling is religiously motivated, but homeschooling parents can also be progressives who reject the rigidity and lack of creativity of organized schools. They may not be able to afford Montessori or Waldorf, or they live in areas without suitable progressive schools. They want to teach their kids to think outside the box, stressing art, philosophy, and music. They may place heavy emphasis on literary interpretation and on writing, fearing that organized school curricula destroy kids' passions for learning and especially for books. The religious right and progressives may meet on a common ground off the educational grid, sharing some of the same criticisms of the one-size-fits all approach of organized schools and wanting to involve parents in the education of their kids.

Prolonging Childhood

Above all, homeschoolers want to slow down growing up, fearing that kids live in a pornographic public culture saturated with sex, alcohol, drugs—mature themes. They may worry about peer pressure, especially when that pressure is transmitted along the electronic nerve system of smartphones and the Internet. As we (Agger and Shelton 2007) argued, childhood has been attenuated in fast capitalism as parents pass along performative pressure to their children—pressure to get grades and test scores and to assemble a college-bound portfolio including extracurricular activities. Homeschooling may reduce performative expectations, even as it denies children certain socialization experiences such as getting along with peers and sometimes hectoring adults. Perhaps homeschooling parents put less emphasis on collegiate and adult success, or perhaps they think that the costs of organized schooling outweigh the benefits, given the acceleration of childhood.

When kids confront homework, even in first grade, and mountains of homework in junior high and high schools, it is easy to see why parents might worry about efficient use of the school day. Many homeschooled kids spend less time on schooling, but still accomplish studying and homework, perhaps suggesting that the school day in organized schools could be laid out more efficiently. Again, costs and benefits: To homeschool their kids, parents, who double as teachers, must sacrifice their own time in teaching and supervising their offspring. And few parents possess the expertise of advanced high school teachers in technical subjects such as math and science, falling back on the omnipresent workbooks and study guides that homeschooled kids pore over.

From Here to There: A Plan for Change

The analysis presented leads to a host of policy, institutional, curricular, and parenting changes. This list could be extended. A critical analysis of society includes a critical perspective on education, especially on schoolwork, teaching, writing. By now, there is a voluminous literature on what is wrong with schooling in a capitalist America (Bowles and Gintis 1976; Giroux 1988; Aronowitz 2008). I share much of those critiques, which frame a specific agenda for change. Such an agenda is meant to start and not end discussions. But the fact remains that schooling in America has, by and large, failed to capture the imagination of young people who frequently hate going to school and thus view education instrumentally, as a means to an end, including pleasing their parents. Kids hide phones in order to text and Facebook because they aren't turned on by what is going on in the classroom. It is difficult to imagine a more significant social problem of the twenty-first century, which is increasingly marked by global competition.

Dialectical analysis demonstrates that oppression breeds revolt and innovation; this is central to Marx's paradigm of social change. I acknowledge that childhood and schooling have become contested terrains. In my hometown of Eugene, Oregon, a progressive college town, public educators and parents push for less structure and more thinking outside the box. Montessori and Waldforf private education is thus implanted in public schooling. My former junior high, Roosevelt, has pass/fail grading. But there is also a nationwide counterrevolution against the sixties, which has gathered momentum in educational policy and practice. My new home state of Texas leads the way in standardized testing, the erosion of evolution in the curriculum, classroom and school building discipline and dress codes, and overall conformity and coercion. Progressive innovations in liberal college towns is matched, and countered, by mainstream schools' redoubling of control, coordination, and coercion and by the growing homeschooling movement, which represents a flight from sociocultural and racial diversity and from pointy-headed secularism. These are not settled issues.

Here are some suggestions for making education more humane and nurturing for kids:

a. View and treat children as citizens with rights.
b. Give students at least an hour of exercise.
c. Deemphasize grading and testing.
d. Re-think the test-score and GPA-driven college admission process.
e. Train teachers in their discipline.
f. Reduce homework.
g. Replace factoids with big-picture knowledge.
h. Integrate posttextual vehicles of knowledge.
i. Promote "secret writing."
j. View and treat attentional disorders as a response to stress, boredom, and lack of physical activity.

k. Open up the classroom and encourage wandering, standing, lying down.
l. Model good lifelong eating in the cafeteria.
m. Decriminalize students and rethink discipline.
n. Replace memorization with gestalt learning and direct experience.
o. Encourage free-form writing and keep kids reading.
p. Compartmentalize mathematics.
q. Start school later in the morning.
r. Put schools into the public sphere and get adults involved.
s. Pay K–12 teachers as much as college professors.

Kids connect and communicate in order to forestall the boredom of school and to share their anxiety and apathy. In the next chapter, I examine Internet postings as cries for help, which they can be. I conclude with desiderata for caring, empathetic adults.

PART III

WRITING TOWARD IDENTITY AND DEMOCRACY

CHAPTER 11
TEXT MESSAGES IN A BOTTLE

Recent studies show that many young people use social-networking technologies to announce their own angst (Moreno, Parks, et al. 2009; Moreno, Vanderstoep, et al. 2009). They also use them to vent their secret desires and to describe their borderline and potentially dangerous behavior. Intriguingly, one author, a medical doctor specializing in the treatment of adolescents, went online, established her own Facebook page, and gently advised kids who "overshared" about their sexual, drinking, and drug exploits to reconsider these full disclosures. Many kids so advised made changes to their behavior, engaging in greater concealment and less online acting out. Other studies (Hinduja and Patchin 2008, 2012; Patchin and Hinduja 2010) de-pathologize kids' social networking use and argue that there is relatively little stalking, bullying, and threats of violence. I suspect that both perspectives are accurate and capture important issues involved in social networking. As with everything technological, there are coexisting upsides and downsides.

I have emphasized the largely positive aspects of kids' online writing, posting, messaging, texting. But just as kids seek utopia online—craving brotherhood and relief from adults' hyper worlds—they also go online to share their pain, even to threaten suicide. This is perhaps not surprising, given the way in which the computer screen provides them a measure of anonymity and takes them out of real time and into the mediated communication afforded by texting and posting. These are two sides of the same coin: Kids seek relief just as they prefigure a better world in their postings. They complain, even in despair, as they reach out to others and seek community. They want others to share their pain and thus to reduce it. Whether or not they are actually self-destructive is beside the point; crying for help already puts them into a significant risk category.

This is not to suggest that text messages in a bottle, tossed out to sea in hope of discovery, are immediately therapeutic. Indeed, pain can be reinforced as kids share their pain, thus perhaps deepening it as they discover that others have found no relief. There is a thin boundary between the pain that kids share electronically and their formation of therapeutic countercommunities that nurture them and buffer their anxieties. If this talk is prison code, as I maintain, certainly the inmates share their anger, aloneness, alienation. The main topic of their complaints is the adult world.

Thematic adults include parents and teachers. Millions of words are written every day by kids complaining about the restrictions and obtuseness of their parents. This venting is good; it is letting off steam. I would have loved to have been able to complain about my overbearing father and boring high school English teacher to such a wide audience! I did complain, but face-to-face, slowly, not quickly and globally. This cuts both ways: In letting off steam to my small group of intimates, I didn't risk turning these passing complaints into global issues. I didn't overshare. On the other hand, I did not receive global feedback from a larger circle than my intimates; I didn't receive the benefit of wider counsel. That might have been a good thing, defusing hot-button issues before they took on lives of their own.

Kids share about the adults hemming them in and just generally not understanding them. They also share about their own feelings of lack of self-worth. They may even threaten to harm themselves. To them, prison—adolescence by another name—might seem like a life sentence, and their problems too overwhelming to endure for long. In this context, suicide seems like an option, especially where they notice that this strategy has entered the electronic discourse of Facebook. Suicide, at least as a rhetorical strategy, has become commonplace; it has been normalized, which, again, cuts both ways. On the one hand, talking about suicide can normalize and hence hasten the act itself. On the other hand, talking about the problems driving one toward suicide might serve as an escape valve for troubled teens who feel all alone. Social networking makes it difficult to feel totally alone, especially as one accumulates a lengthening list of network contacts who are privy to one's private musings. Some ten thousand Americans die in gun homicides every year, but over twenty thousand commit suicide using firearms. Although school shootings like that in Newtown, Connecticut, are horrific, approximately seventy Americans take their own lives using firearms each day, an issue treated in my forthcoming coedited book, *Gun Violence and Public Life* (Agger and Luke, forthcoming).

Moreno, cited above, is conducting research on whether troubled postings constitute a cry for help. Psychologists have often urged parents and educators to take notes threatening suicide seriously. Even if suicide is only a rhetorical move, this is not a move that kids take lightly, given the stigma attached to it. Although they may be manipulative, calling attention to themselves, adults cannot afford to ignore these cries for help—even if the help is not directly suicide prevention but a questioning about why kids would need to manipulate the impressions of their online friends. Why aren't they getting what they need? Indeed, that is perhaps the central question that adults should be asking as kids pour out torrents of words concealed in the casual code of the Facebook generation.

In reading social-networking postings, I am struck by both what Gould calls oversharing—the revelation of very intimate details to a large circle of friends and even strangers—and by the rhetorical excesses often in play. Perhaps adolescence is, by definition, exemplified by the lack of a sense of proportion. Things are felt and expressed dramatically: how Johnny dissed Susie; how one's parents are mean; how crazy that party last Friday was. The fact that kids experience these slings and arrows as cause for writing perhaps reflects both their stage of development and their access to means of self-expression. Students of childhood eating and obesity suggest that kids are overweight and out of shape because food is everywhere, readily available. By the same token, Twitter invites the trivialization of discourse as people simply track their own movements during the day. By definition, there can be a limited audience for such ephemeral jottings.

Oversharing

Kids seek a public. They are in the process of developing what Cooley called the looking-glass self—the adult person who can reliably imagine how others view him or her. This public helps the self create himself, mirroring a growing sense of identity. "Am I nuts?" "Do you agree with me?" The young person is testing reality. To do this requires that one disclose a lot of information, especially because responses can be random and not composed and delivered in real time. Writers will repeat themselves and miss connections. Perhaps oversharing is best understood as the ratio of messages sent to messages received and responded to. It is a sheer function of the volume of communication required for the young self to measure himself or herself, in Cooley's sense.

As well, oversharing involves a certain sensationalism, a self-sensationalizing. The boundary between truth and fiction fades as kids become their own heroes. Although Twitter embraces the quotidian and ordinary as important texts, Facebook seems to provoke grander narratives, sometimes in full color and with a musical score. Images and graphic design become texts in their own right, as does music. This makes sense in that young selves are iPodified and iPhoned; they are defined by their music and their taste in videos, movies, and television.

As well, the social networking services provide questionnaires that lead to a great deal of self-revelation, all the way from relationship status ("single," "it's complicated") to favorite foods. Some ask the young person to comment on the quality of their parents' relationship. Simply posing the question that way leads to sarcastic and potentially problematic responses, where not only is too much disclosed but there is a great deal of spin and exaggeration. And how many teenagers really "get" their parents and their relationship? The gap between the generations works both ways.

Cries for Help

Oversharing may occur for a reason, not simply to make the private self public in a self-revealing, narcissistic way. Young people, when troubled, may take to the airwaves to share their pain, even to threaten to harm themselves. Code busting is not easy, any more than it is easy for the trained therapist when counseling a depressed patient. While I was finalizing this book, the elementary-school shooting took place in Newtown, killing twenty young children and a half dozen adults. Initially, the shooter was misidentified as his brother, Ryan Lanza. But the younger brother, Adam, was the disturbed gunman, who took his own life after the school shooting and after killing his mother. The misidentified Ryan Lanza took to Facebook in real time to express his innocence. He was understandably troubled and intense, and screenshots of his Facebook page soon appeared on the Internet, as did his photo. It is inconceivable that he won't be scarred by this tragedy, of which his misidentification was just one part, but a significant one. The Facebook screenshot helped this misidentification go viral, but it also gave the real Ryan a vehicle for clearing himself. He said, in a poignant way, that he couldn't have been the culprit because he was riding a bus home from work.

Every adolescent and young adult act of madness these days immediately sends forensic experts and psychiatrists back to the electronic records that transcribe anxiety and even insanity. Parents worry that their own kids are baring their troubled souls in their texts, posts, and blogs. It takes self-discipline not to hover and snoop. After all, kids need privacy, a life and room of their own. But savvy parents recognize that kids are writing their lives on the electronic screen; pixels become autobiography. Kids who are depressed, sullen, anomic, or anxious post their troubles on Facebook, and they text their troubles to friends. "I hate school" is a common theme, as I discussed in the previous chapter. Other adults, including parents, come in for comment—as do boyfriends, girlfriends, and exes.

Letting off steam may not be a cry for help. And kids, like their parents, spin and even lie. But there is truth even in falsehood and exaggeration. There may be a thin boundary between therapeutic and needy posts and texts; announcing one's blue mood could ease the way or signal a serious downward spiral. As we have learned from theorists of interpretation, from Nietzsche and Wittgenstein to Derrida, context is everything. Posting after the dissolution of a high school relationship may be quite different from posting after one gets a C in algebra.

The Emphatic Self

Social networking posting is perfect for manufacturing drama—the thick text of adolescent lives. People of any age can use the Internet for self-invention, even for self-inflation. Young people can find trouble where none exists, and they can problematize their own lives and the lives of their peers simply to get attention. This is the emphatic self, with emphasis added. There may be a flimsy boundary between lying and adding emphasis. That boundary is perhaps where drama begins. And drama is best understood as the need to entangle others in one's own web, one's own selfhood. Gossip has always lubricated social relations and sometimes thwarted them. Electronic self-expression, given its economy and instantaneity, is especially suited to gossip, understood as the drama created by selves who exaggerate in order to test their own efficacy as social actors.

This can take the form of passing along tidbits of information, especially about romantic and sexual relations. Or it can take the form of public nastiness about others, whether peers or adults. It is easy to see how this can lead to bullying, a curious outcome of the transmission and exchange of pixelated public and private discourse. Adults do the same things, especially in the workplace, as they transact relational power, including proximity to powerful people who might prove advantageous to one's careers. Adults often deploy email for this, while kids text, tweet, and post. And adults dexterously and sometimes disastrously use the cc or bcc functions in order to create networks of drama.

That social relations can be distorted quickly, in real time, using electronic communication with global reach is a peculiar feature of a postmodern world. However, our world is still modern—pre-postmodern, in a word—in that we still possess standards of truth and falsehood with which to assess the claims people make, text, post. Drama can be measured against reality, even if reality is open to multiple interpretations. That Mary cheated on Johnny by kissing Billy while drunk at a party can be verified, even though it may not have been a deep and meaningful kiss and even though having a beer may not constitute inebriation. Perhaps Billy was not even present, or lips never touched. The ease and speed of electronic communication lead to almost automatic reification of these claims; that is, they take on lives of their own as people take them to be true. As well, the gossiper—the writer, as it were—cannot be viewed as free of motives, whether or not she lies or spins. Drama is created out of need, even, no especially, if the need is not transparent to the person telling the tales.

Perhaps the most notable thing about adolescent electronic writing is the need for emphasis, for exaggeration, for drama. "OMG!" Maybe adolescence is best understood as the best and worst of times, a poignant passage between the carefree elementary years and the relative sanity of young and older adulthood. Perhaps adolescence has always been about drama and the need for added emphasis. Slights and small victories are more acutely experienced than earlier or later in life. So much seems to be resting on how one's friends view you, on one's status, on one's group affiliation, on one's perceived sexuality. The adolescent imagines that everyone is staring and judging but also contributes to that awkwardness by oversharing, and in ways that sometimes defy credulity.

In this sense, the normal pains and pangs of adolescence are intensified by the electronic public sphere that makes oversharing inevitable. Although some kids resist the lure of social network sites, many others view them as daily journals on which they imprint themselves. When I was a kid, no one read my scribbling, unless my mother found my letters under my bed. Today, the whole world can gain access to the emphatic self, the self who, by exaggerating, adds significance to his or her own life. As before, adolescents lack self-esteem and now adults do, too. Social network posting is a quest for identity, for value, for place.

Simultaneously, kids can now reach out and build community electronically, just as they can hide behind the screen, enjoying a social distance that frees them from the emotional risks of intimacy. It is well understood that one can say things on the screen and via texting that one wouldn't say face-to-face, for fear of embarrassment. This is good where people, especially adolescents, may be shy. It is potentially problematic for kids who produce drama by revealing an interior life publicly, where before they buried these thoughts in a diary. This is not to presume that electronic postings, messages, and texts accurately reflect the inner adolescent self. They may exaggerate or even falsify—precisely the nature of drama.

If stress is the postmodern form of alienation (Marx's nineteenth-century concept, closely allied with Durkheim's notion of anomie), drama is a way to deal with stress, even if it reproduces it. Drama involves gossip, self-revelation, hyperbole, and sometimes sheer fiction. It positions the self in the center of life's script, especially where kids acutely experience their own marginality and lack of civil rights. Stress and accompanying drama are not the preserve of the young; adults engage in the same sort of self-inflation in order to combat their deflation by the swirling social forces surrounding them. We are all living damaged lives, to use Adorno's term.

Major shifts are taking place in modern America, and beyond our borders, as the Internet enmeshes us. We spend more time online. Our postings, messages, and email reproduce themselves, with responses prompting responses in a cybercircuit between senders. Older people often find all this to be a burden, with email and messages weighing them down. Younger people learn how to deal with this dexterously; after all, they never knew a slower world in which people simply walked down the block and knocked on doors in order to make contact.

Self-revelation is one of the consequences of the shift from neighborhood to network. Twitter takes this to the ultimate degree: People can track us electronically throughout the day, as we perform one mundane task after another and record them as tweets. Writing and everyday life blur as our mobile phones double as word processors. Traditional writing was crafted in the lonely writer's garret, at a remove from the world. Sentences were considered carefully, revised, polished. Now, the words pour forth as kids and adults write on the fly, as they sit in waiting rooms, wander off to class, even while driving. They are limited by the number of characters their phones allow, or perhaps they aren't limited at all, if they blog. But their writing is awash in instantaneity, as they write without much reflection. This is not to denigrate. Fast writing can be captivating and charming, and slow writing can be leaden. Twitter is a vehicle of instant expertise in this Age of Opinion, as people, in real time, record their reactions to the world around them, contributing to the cacophony of a cyber-public sphere.

Two interrelated questions present themselves. Both involve boundaries. What is the boundary between "real" writing and electronic writing? What is the boundary between writing and living? From my perspective, what we consider "real" writing is composed carefully, at a remove from everyday life. It may proceed from an outline and involve numerous self-edits. Electronic writing, although it can be distant from everyday life and composed carefully, is less argument than a signaling, an electronic wave that says "hey, I'm over here"—almost more like oral communication than writing. It need not be this way. After all, I am composing this book on a laptop computer. I could conceivably compose the book on my cell phone and send it to the computer through a series of text messages. That would take a lot of patience, but it could be done. Even to suggest the distinction between arguing and signaling—presenting a considered case and saying "look at me"—neglects the millions of messages, emails, postings, tweets that convey information and even present reasoned argument. And one can compose the old-fashioned way, apparently from a distance and with an old-fashioned writing instrument, and in effect engage in signaling—a love note exchanged between teenagers, for example.

My argument is that the boundary *can* be breached as everyday writing takes on the considered and critical characteristics of studied argument. Husserl, the phenomenological philosopher, distinguished between life lived in the "natural" attitude and life lived, from a distance, in the "theoretical" attitude. By this he meant that people, when they cross the street, don't consult a manual explaining what their behavior should be or delving into the chemical composition of the painted lines on the street. Only when they remove themselves to the Apollonian peaks of theory do people interrogate taken-for-granted everyday assumptions; they philosophize.

Husserl's phenomenology drew attention to the importance of everyday life—here, what kids write and how they communicate with each other. But he was destining everyday life to be a site of pre-theoretical, unreflected behavior; he was not a utopian who felt that people, once liberated from ideologies and false consciousness, could elevate themselves and theorize their lives, eventually easing the boundary between composing a considered text and tweeting.

Husserl influenced later developments in ethnomethodology and social phe-nomenology, currents in 1960s and post-1960s sociology that view people as capable citizens and not simply mute Parsonian role players. The 1960s insisted that important social changes should flow through and transform everyday life—between the races, classes, sexes, nations, and cultures. The civil rights movement, antiwar movement, women's movement, and counterculture took a bottom-up, grassroots approach to change, building massive social movements out of the individual protests and reforms by college students and other youngsters who, as Mario Savio urged, put their bodies on the gears of the machine and forced it to a halt or, alternatively, loved one another right now, as The Youngbloods recommended.

Just as we must interrogate the boundary between "real" writing and the electronic kind, especially as we develop a utopian concept of an electronic pub-lic sphere in which everyone has a voice, so we must question whether there is a clear boundary between writing and living. Digital communication connects people and produces a literary instantaneity; we can respond to anyone and anything in a public way as we are on the go, driving or being driven, at work, in class, anywhere. The image of the writer's garret fades as writing takes place amid an everyday life which is perhaps no longer clearly differentiable from the extraordinary lives of writers and intellectuals.

On the one hand, this sinks writing and reflection into the maelstrom of an unconsidered everyday life—precisely the image called up by tweeting cor-respondents who simply report on their whereabouts and activities: "I'm at the gym and afterward I'm getting my nails done." Or the banal texts that report the same things, but directed to another correspondent. It is writing only in the thinnest sense. On the other hand, this suggests a literary utopia in which citizens post and publicize as they bridge the natural and theoretical attitudes, as Husserl termed them.

It is difficult to avoid banality when texting and tweeting. These formats were made for simple information transfers and jokes. However, other rapid information and communication genres such as blogs open writing to the sort of essaying that constitutes the sinews of a democratic polity. As I noted earlier, to *essay* is literally to *attempt*. To attempt what? To attempt a life, a public life, and to build like-minded community. Young people seek this community, even as they have insufficient intellectual tools to achieve it, mired as deep as they are in everyday life. Writing, for kids and also for adults, becomes simply mimetic, reproducing and repeating the jargons of inauthenticity enveloping us.

The flimsy boundary between writing and living is pierced as kids write their lives, constituting their social relations electronically. This is pre-political, auguring a future world in which citizens create and control their worlds. It is not yet political because texting and tweeting kids don't intend their electronic connections as transforming; they are merely using the expressive codes of the moment to laugh out loud at the adult world. This was much the stance of fifties and sixties rock music. Kids sang and danced a new world, but without a heavy theoretical apparatus that used terms such as the "performance principle" to understand the world from which they were trying to break. Only when rock bands staged benefit concerts supporting the antiwar movement or progressive political candidates did culture bleed into politics. I kissed a girl at a Grateful Dead concert in 1969 that was sponsored by the SDS. Neither the concert nor the kiss was political, but, then again, I was already learning to be free of the straitjacket of mainstream adult society that perpetrated Jim Crow segregation, poverty, and the war in Vietnam. I sensed, as kids do today, that they can pre-figure new worlds characterized by companionate and egalitarian relationships held together by the nearly instantaneous electronic technologies tethering kids in a vast body politic.

This is not to portray children and teenagers as utopian figures, free of the soiling influences of civilization. They are damaged by the world, much as we, their parents, are. Electronic chatter and postings are often cries for help, signals that kids are in trouble and possibly alone. What might seem like self-referential narcissism in postings by the young, who disclose their innermost turmoil and wounds—what I call *disturbia*—are in fact messages in a bottle destined for caring souls who might ease their burden. Kids painfully grow egos and grow apart from parents, confronting their sexuality and reckoning with a modicum of freedom. Today this is arguably more challenging than ever because adults are nearly totally absent from the latchkey upbringing of kids, as busy parents deal with work and their own relationship issues. In the recombinant families torn apart and stitched together by divorce and other social pressures, everyone is needy and, almost by definition, no one is having his or her needs adequately met.

Part of this involves time, in our fast and faster capitalism. Time is sucked up by television, the Internet, mobile phones, laptops. Work invades the home, and the family is "done" in the workplace, especially by women. People are always "on," and on call. There is little downtime and scanty sleep. Kids claim the night as the time when they make electronic connections, after (or instead of) homework when the parents are asleep or out of sight.

This appropriation of the night has a utopian dimension: Kids are living as if they were free, unbound to adult authority and routine. Yet, being untutored in utopian theory, few kids theorize their nocturnal electronic emissions as the message in the bottle from the shipwrecked, let alone as a utopian prefiguring. Sometimes parents read their kids' nighttime postings with alarm, as the disturbed and sometimes brazen expressions of youth gone bad. They cannot believe that *their* kids have gone so wrong. Kids intend to be given an alarmist reading; they want to hide behind the computer screen, trying on and posturing identities somehow impermissible in everyday face-to-face interaction. And adults do this, too, writing themselves in ways that are often idealizing and always distorting.

This is not to trade on a stable concept of the "real" self, especially the adolescent self. The very challenge of adolescence is to arrive at internal integration, staving off the persistent impulses from libido and rage. Freud characterized this as growing or developing an ego that would mediate between infantile desire and morality (id and superego, respectively). In a sense, then, kids' nocturnal writings can be read as the work of the unconscious, much as Freud attempted to gain access to repressed elements of the adult self by way of dream analysis. Electronic writing exists in a space between dreams and waking consciousness; it flows freely from fingers that type and text faster than the conscious mind can mediate them. Perhaps all writing proceeds this way—with serendipity as much as the structured plan laid out in the outline or table of contents. Good writers allow the writing to write itself, learning from the flow of words and allowing that flow to revise the outline as the literary logic of the text unfolds.

And so kids' possibly disturbed and fantastic writing needs to be read symptomatically, as perhaps not fully formed expressions of an unconscious working itself out on the screen. In this sense, " badass" might be read as "rebellious" and "I want to end it all" as "I'm looking for a friend and a community." These symptomatic readings cannot risk being wrong. Seung-Hui Cho, the shooter at Virginia Tech, littered his college English papers and poems with evidence of his disturbed state of mind. Two caring English professors at Virginia Tech read through his rants to a potentially destructive and self-destructive intentionality; they tried to help, but he was already lost to their interventions in the impersonal association of the college campus.

Decoding the electronic talk of the young requires interpretive skill, both because the Internet and cell phones facilitate a *practiced indirection* and sometimes sheer deception and because there is a tendency for adults to read kids' talk *out of context*, or perhaps better, context. The shielding afforded by the screen encourages kids to be other-than-themselves, although Freud would read these tracings as highly relevant, perhaps stemming from an unconscious desire to be found out. As well, there are few social costs to be borne by the posturing, even deceiving author; the electronic talk of the young is frequently shameless as the young and even their parents bare their souls but don't risk blushing or being physically shunned. Adult readers frequently lack context within which to situate the quick texts that may spring from the sort of practiced indirection I am talking about. Seung-Hui Cho left many face-to-face and electronic traces of his growing derangement. Even then, full context was never fully grasped by the surrounding adults, allowing him to move beyond disturbia to full possession by his demons.

Taken out of context and heedless of the postmodern plasticity of the electronic adolescent self, adults can read disturbing postings and transmissions as more meaningful than they are. At the very least, they will be shocked by the overtly sexual content. Or they can read these as less meaningful than they are, as "kids being kids." We need interpretive strategies—call them sociologies—that make sense of these insubstantial but potentially important writings. These would be e-sociologies that could be practiced in everyday life, weighing text and context in order to make sense of the lives our kids lead. These are the ways in which we read our kids as we struggle to understand the worlds we have imposed on them. It is not enough to be disturbed by their messages; we must empathize with them, wondering why they are trying to claim the night as their own and evaluating whether there are utopian impulses in their efforts to write their own worlds.

These strategies of empathetic interpretation (again, sociologies, on a certain meaning) require adults to make these commitments:

- We will treat kids' writing as serious outcomes of deliberate authorship.
- We will wonder and worry whether our kids are crying for help.
- We will allow them space and time to write and live below, or beyond, adult radar (with our "performativist" expectations about achievement and success).
- We will show our kids that we share many of their concerns and that we sometimes have trouble making and sustaining connections.
- We will try to establish context for what we read and not overreact.
- We will show our kids examples of our own earlier activism as we model a cyberdemocracy that carries forward sixties democratic social movements.

Everyone talks about the obesity epidemic, especially among the young. In the next chapter, I consider the alienation of bodies in the context of a positivist belief system that reduces health to simple, single numbers, such as weight and BMI. I suggest that we should emphasize not bodies of knowledge but rather knowledge of the "body subject." Such knowledge is gained through exercise conceived as play, self-expression, and self-awareness.

CHAPTER 12
BODIES OF KNOWLEDGE

Kids become so adept at using their phones to type and browse that we notice that mind, body, and voice blur to the point of near identity. At the same time, plugged-in kids, linked in chatter and community not available on the adult wavelength, don't use their bodies in locomotion, the way we did. They come to hate their bodies as these young bodies don't measure up to cultural ideals of youthful beauty. Another way of saying this is that writers must make an effort to exercise.

Haruki Murakami (2009), noted Japanese novelist, presents us with a memoir more about his running than his writing. But clearly the two activities inform each other as he learns the world by (and by way of) learning his body in motion. I am especially interested in young people's alienation from their bodies and minds, an alienation expressed as low self-esteem. Schooling, popular culture, advertising, fast-food diets all estrange kids from their impulse to play—with their bodies and ideas. Although I view fitness and health as obviously important, I am also interested in "working out" as an antidote to alienated work (and homework), indeed, as an expression of freedom and appropriation of meaning. Running is hard work, and good for bodies, but it is also purposeless, an activity, like any form of play, embraced for its own sake. This is exactly what early Marx termed *praxis*—self-creative activity performed beyond the cash nexus and performance principle.

Images of the alienated body abound, in advertising, entertainment, even science. Empiricists often work within "bodies of knowledge," to which they make their small contributions. This derives from an alienated body politics, which, according to early Marx, lies at the heart of the logic of capital and its manifestation in the wage relationship. For him, capitalism alienated bodies, labor, community, nature, even (the Frankfurt School adds) science. In this sense, I am interested in how positivism, a moment of alienated body politics, produces body sciences, including exercise physiology and nutrition, that are themselves alienations, perpetuating the duality of self and body that is paradigmatic of all alienation. I also work toward a critical theory of bodies in motion, including bodies of knowledge, that moves beyond critique.

Positivism is an alienation that borrows from the discourse of embodiment—for example, the term "body of knowledge" referring to what Kuhn called normal science. This body hangs together skeletally, according to the literature-review sections that begin standard empirical journal articles, through the artifice of the author, who claims, in parenthetical string citations, that there is already consensus on fundamental findings. The rhetorical art of the lit-review section is at once to compose the body (of findings) and to propose the novelty of one's own research, hence allowing the body to grow, evolve, adapt. Only by accomplishing this sleight of hand, can the busy article author seek tenure and promotion, or even the first academic job.

I have written about how this approach to science necessarily narrows knowledge, hence contributing to the decline of discourse, theory, books themselves (Agger 1989b). This approach to the positivist body of knowledge belongs to a larger category of what I call "alienated body politics" that stretches from the male gaze distorting women's bodies to considerations about food and exercise. Positivism, a theory of knowledge, is only one manifestation of alienated body politics. These alienations begin with capitalism (when alienation of labor first emerges as labor, and hence the body, is commodified) but predate capitalism, especially where religious traditions put a hex on the sinful body and then, with Descartes, split it from the elevating mind.

I seek a way of talking about (a "discourse" of) the relationship between the body and the world that reverses (disalienates) the body. Instead of bodies of knowledge, I seek knowledge of bodies, but not simply an objectivist knowledge reduced to quantitative indicators (e.g., weight, blood pressure). This is self-knowledge, the kind gained through a unified/unifying mind-body that experiences its own metabolism with the world and nature. I contend that this unified/unifying experience of the self is the organon of disalienation, of embodied freedom. It is a utopian moment at a time when utopia has been suppressed (Jacoby 2005).

This is risky because exercise and food quickly become occasions of self-absorption for yuppies who care little about alienation and who thrive within capitalism. Literature about running, spanning the first running "revolution" (1970s–1980s) to the present "second" revolution, segues from the edgy and even political (Sheehan 1978; Henderson 2004) early issues of *Runner's World* to the current conformist and commodified version of the magazine. Henderson no longer writes for the magazine, and it recently featured a celebratory spread about Sarah Palin. Body talk can easily lose sight of intersubjectivity, awash in its own expressive subjectivity, a topic treated by Jacoby (1975) in his discussion of the "politics of subjectivity." Within a narcissistic culture, bodies can seemingly be healed without a general healing (a therapeutic synonym for socialism, perhaps).

I do not think it is that simple, although change has to start somewhere—change here referring to a disalienated body politics. Small steps: exercise (as play), grow and eat healthy food, build like-minded community, and eschew fast-capitalist fixes such as drug-oriented medicine, crash diets, stiletto heels, steroids, and plastic surgery.

Without using metaphors that unintentionally narrow, running might be an example of the playful subject-object (unified mind-body) who develops knowledge of the body that is at once self-knowledge. When one is in motion (and it doesn't have to be running), one attains a state of unified subjectivity/objectivity that refuses to—or simply cannot—distinguish between the mind and the body. They are one. This happens rarely for me, and I've been running for over thirty years! Much of the time I am not in "the zone," which is just the moment of unification I have been talking about. Regularly, the body feels like a drag on me: it might be sore, the motion feeling like work. With age, one can be fit and still experience the body this way. But one runs for moments of unity, of flow, that afford clarity about the unified self. This is perhaps what utopia feels like.

This is the kind of thing George Sheehan was talking about in his running books, including *Running and Being* (1978). Sheehan was lampooned for investing too much philosophically in running. But a careful reading reveals that he was talking about these issues of mind-body unity as a reversal of an alienated body politics. His progressive politics were often overlooked by yuppie runners who pretended that exercise was value-free, a version of bodies in motion that drew heavily from positivist assumptions about how the self is outside the world and can be known itself only from the outside. Self-knowledge is derived from bodies in motion. Gunnar Borg (1998) demonstrates that ratings of perceived exertion correlate strongly with actual exertion. If it feels difficult, it is difficult, ascending that hill or sprinting in the last half mile of a race.

Exercise is related to food as motion is to fuel. Carbohydrates and even fat produce energy—sugar converted into glycogen with the aid of inhaled oxygen. Aerobic exercise has enough oxygen to effect this conversion, and anaerobic exercise lacks adequate oxygen. One might observe that both are necessary for self-knowledge.

Michael Pollan (2008) argues, against "nutritionism," that food cannot be known only from the outside, or working from the outside in by tallying the nutrients, vitamins, and calories in food. So-called primitive cultures knew what was good for them, which was usually equivalent to what they could find in their native environments. These were diets heavy on complex carbs, plants, lean meat, or no meat. Most of these people, such as the Mexican Tarahumara, walked and ran long distances for their food, enhancing their fitness by chasing their food ("persistence hunting"). In these cultures, mind and body go undivided; they are pre-Cartesian cultures, which may be another way of describing them as pre-modern. I consider the shift from pre-Cartesian/pre-modern to Cartesian, modern societies as a source of numerous "body problems" that bear certain obvious physical costs such as high blood pressure, heart disease, and obesity but also some less-obvious psychic costs such as distance from one's body and one's life (Agger 2010). Young people's alienation involves schooling, adults, food, and the absence of exercise. It forms a totality, as all alienation does. An implication is that band-aids won't fix it; we must rethink everything, from sleep deprivation to smartphones.

Theses

All of what I've talked about thus far can be summarized as theses:

1. Capitalism makes us sick and then attempts to heal us with commodified fixes such as weight-loss diets, supplements, gymnasia memberships, even plastic surgery. Capitalism harms us because alienation affects bodies and then tries to heal us.
2. Earlier, bodies under Fordism and pre-Fordism labored under cruel conditions.

3. Now, bodies under post-Fordism don't labor enough but are squeezed into cubicles and school desks.

4. Diets grounded in processed food, factory meat, and high-fructose corn syrup (Fordism applied in agribusiness) contribute to coronary artery disease, high blood pressure, weight gain, and the atrophy of the joints and connective tissues.

5. Positivism, which is an instance of an overall alienated body politics, pretends that we can study the body objectively, from the outside, using various health indicators measured at the annual physical exam. These indicators, such as body mass index (BMI), are almost always quantitative.

6. The BMI ignores percentage of body fat and hence discriminates against athletes, thus rewarding people who are sedentary. LeBron James is thus classified as overweight, which is absurd.

7. What Pollan calls "nutritionism" examines food in terms of its chemical and nutritional constituents and then "enriches" food that has already been processed. There is little evidence that processed food, once enriched, is equivalent in healthfulness to the original unprocessed food.

8. The mainstream indicators of health revolve around weight, which is grounded in versions of acceptable femininity and masculinity. There is little evidence (see Oliver's *Fat Politics* [2006]) that obesity can kill; instead obesity is one of the side effects of the fat-laden standard American diet, eaten by largely sedentary people. Other more dangerous side effects include high cholesterol and hypertension.

9. Weight is a convenient positivist obsession because it is a single number, because the diet industry is lucrative, and because it rewards women for starving themselves, only lowering their self-esteem.

10. A nation that exercises necessarily reduces time spent in paid labor and in productive consumption.

11. To be healthy one must be fit (Emerson's "good animal"), but fit people can be unhealthy if they are alienated from their bodies and food.

Science is not a body (of findings) but an embodiment, a mode of being in the world. A disalienated version of science suggests science as play and praxis, like painting a picture or going for a run. A Cartesian version of science splits off science from the world and creates a "body of knowledge" when in reality science is already in the world and cannot escape the gravitational pull of time, place, body. First Einstein and then postmodernists jettison Newton's physics, which pretended to stand outside of history, the world, the body, especially using the distancing technologies of mathematics. This was positivism's only apparent escape from the prison house of language, and it fails because science is itself rhetoric—a way of making an argument, even where it is secret writing (the author disguised under the heavy layers of "objectivity").

It is no wonder that we probe bodies in reducing "health" to various indicators such as BMI and cholesterol count, just as the literature review describing bodies of knowledge flows into the quantitative segment of social science journal articles. Bodies, reduced to inert lifelessness, cannot be bypassed by Cartesians. Embodiment is not to be shunned but embraced, even as bodies age and slow. Bodies (of science and in motion) make themselves available to be known, and improved, through a deep self-knowledge that does not rely on the distancing techniques of method to keep perspective, passion, and politics at bay. To note that language is a prison in which meaning is incarcerated does not condemn language but requires that truth—truth*s*—are possible within writing, but not of the kind that pretends merely to describe. Even description advocates, the more it appears not to take sides. I call this the "secret writing" of positivism. By the same token, we cannot summarize health in body indicators. Instead, we must eat and run like the Tarahumara—swift and enduring utopians.

McDougall's *Born to Run* (2009) chronicles the running lives of the Tarahuma who, in his telling, meet and compete with leading American ultramarathon runners. It is clear from his account that the Tarahumara run as a form of play now that they don't stalk their prey. But his tale reveals that ultrarunners (i.e., in races longer than the 26.2-mile marathon) are our utopians: They inhabit a cooperative community in which runners help each other and they run to express themselves and explore their limits.

Turning toward the body might be seen as a departure from politics. Indeed, for many of us who grew up during the 1960s, this is precisely what happened as we experienced the right-wing retrenchment of Nixon, J. Edgar Hoover, Reagan, and the Bushes.

I left the United States physically in the 1960s, along with many other foot soldiers of the New Left. I also retreated from politics to "theory"—an academic life—and running, along with other personal pursuits, including love and eventually family. Was I running away? Or toward? I theorized running as a non-Cartesian merger of mind and body, a connection to mother earth. I was looking for America. My texts were Robert Pirsig's (1974) *Zen and the Art of Motorcycle Maintenance,* a book everyone read. And an obscure classic called *Meditations from the Breakdown Lane*, James Shapiro's (1982) chronicle of his solo run across the country. An apolitical book written by an ex-radical was really very political: like me, Shapiro was looking for a home within, an existentialist reaction to the end of the movement. Shapiro's final words: "The bear went over the mountain to see what he could see. And what did he learn? That everywhere there is sky, everywhere there is ground. At every moment, everywhere, we are home."

Feminism developed its personal politics by addressing home, family, body, sexuality. I developed mine through exercise. Mark Wetmore, the University of Colorado track and distance coach, borrowing from Tom Wolfe, talks of an Edge City of extreme physical exertion—running not into oblivion but into meaning (see Lear 2011). Few young people seek Edge City these days, whether in running or working. They are not to blame; Edge City—another name for utopia—has been malled over. We of the 1960s still search for community, albeit in ways and places uncharted during those original times.

Our kids grow up joined at the proverbial hip to their electronic prostheses that allow them to form community. But their bodies are inert, confined to the school desk and couch, unless they play organized sports (which are usually an alienation because they are bureaucratically organized and run by adults who stress winning). Kids need exercise, but conceptualized as play. They also need cognitive play, not rote learning and fact-based curricula. By attenuating childhood, we suppress children's play impulse and foster depressive mood disorders, malaise. Instead, we need to get their bodies and minds in motion.

In my concluding chapter, I explore life as a journey to nowhere, but not for that reason a journey not worth taking. Young people are adept at being in the moment, and we can learn tranquility and serendipity from them, as well as from journey runners. This is a blending of early Marx, Zen, and running theory. I also suggest ways for adults to nurture children, including valuing their writing in this age of unbound books.

CHAPTER 13
DEMOCRACY IN THE FIRST PERSON

We learn from our kids that literary work is good work, enhances democracy, and makes us human. We might lament their lack of pulpishness, but, truth be told, few adults, even academic ones, write. We assign writing in our classes, but most teachers and professors don't pick up the pen or pound the keyboard. When we do, we tend to compose in our own restricted codes, jargonizing and narrowing our work (Bernstein 1975). We address hundreds of like-minded scholars, not hundreds of thousands of general readers. This story has been told admirably by Russell Jacoby (1987) in *Last Intellectuals*, a story about how sixties intellectual activists decamped to the university after the decade ended and learned to write for tenure, not for the public.

This is a zero-sum age, an age of dichotomies. Derrida, the father of postmodernism, instructs us that dichotomies always contain secret hierarchies, such as men over women. We feel that we cannot have both thick, slow pulpish discourse, requiring monkish reading and reflection, and the hyperreal discourse of children's language games, from texting and tweeting to Facebook. These seem opposite, much as our children seem the opposite of us. We sixties parents are passionate, driven, empathetic, perhaps hubristic. Our kids slack and slouch; they blog instead of write and we judge this inferior.

I confess my prejudices: I want my kids to know who fought whom in World War II. To be unaware ignores the Holocaust, in which some of my relatives perished. One cannot be alive and sensate without understanding that the

Nazis, abetted by the good Germans who turned a blind eye, blended myth and enlightenment—blood pride and the factory-like killing camps. September 11 was terrible, but it has many precedents and on much larger scales. Newtown unleashed murderous rage of the young upon the younger. I want kids to understand themselves via their pasts and the pasts of their forefathers and foremothers. Social amnesia flattens the present into fate, a plenitude of being.

But kids have different, not necessarily inferior, knowledge. They surf, connect, compose, disclose. They do democracy if by that we mean egalitarian social relationships grounded in dialogue and reciprocity. Kids don't hate or see color; they learn to do these things. The digital world lacks hierarchy; it is open to all comers. Slacking is a reasonable response to a frenzied, overburdening world; it is not an embrace of laziness but a questioning about productivism—output for its own sake. Slackers are utopians; they want peace, including the opportunity to be you and me. Where authoritarians have never found a rule they didn't like and seek to enforce, slackers question authority, including the authority of elders.

This is not an endorsement of anti-intellectualism. A world full of music and chatting also requires books and blogs. People need to read and consider arguments carefully. Blogging should be matched by reading, which is the resource of writing. Texts are nucleic societies of readers and writers; they prefigure a wider democracy. This is what Habermas meant by the ideal speech situation. A crucial difference between pre-sixties textuality and the post-sixties kind, that emerges in texting, messaging, and blogging, is that only the latter allows for the presence of the first person, which I read as the basis of humanism and democracy.

Positivists and postmodernists alike reject the first person, the former because the self sullies objective knowledge, the latter because the subject is dead and there are only subject positions. The 1960s installed the self as the basis of its humanism; that is the message delivered in the opening "values" section of the SDS's *Port Huron Statement*. Young people, clinging to subjectivity, insist on their own value, by journaling—blogging by another name. By insisting on the right to compose themselves in the first person, they make way for many persons, for dialogue. Kids' prison talk creates a collective subject; it is protest that prefigures democracy.

Social media brim over with talk of the self. Tweeting is the pixelation of the self; communication technology and the writer achieve full identity. It staggers adults that kids would bother to record their every neural twitch, sexual preference, favorite band. The first person quickly becomes an untamed narcissism. Yet there are blogs that are studied and well versed in affairs of today and yesterday. They are sites of self-expression, intellectual problem solving, dialogue. Blogs publish the self, allowing it to go public, in ways that traditional, especially positivist, academic discourse does not. The blogosphere is the public sphere by another name—itself another name for democracy.

It is clear that commercial and academic publishing are having a difficult time competing with the open-source Internet. One worries about the future of books and libraries—opportunities to compose oneself slowly and to be considered deliberately. Pixels are cheaper than pulp and they zoom globally; books are expensive and inefficient by contrast. Of course, that is why we love them! They take effort, to write, read, and publish. This is the age of effortlessness, not the Edge City imagined by Tom Wolfe where people push themselves to the max in order to find out how much they can take and accomplish. Slackers slouch, even if they also blog and text. It is difficult for overloaded kids to rise in the morning, the time when some creative people do their best work. Many would protest that one can work productively deep into the night, which may be true. Time rebels ignore the clock and follow their muse, making their own schedule. But few would deny that most young people do not take up residence in the Edge Cities of their choice—maniacally improving their cello technique, their writing style, their sports ability—unless they are pushed by parents and coaches, who care less about Edge City—excellent play—than about winning and resume building.

Adults like me may simply miss the huge efforts expended by young people as they write the self, play music, start businesses, build community. We elders tend to judge effort by our own standards of grades achieved, books published, money earned. Such outcomes are not irrelevant in a challenging world. But perhaps the young are more into process, more Zen; in that sense, they are perhaps more like we were when we were reading slacker utopians such as Pirsig and Kesey. We used to be "in the moment," too, enjoying process more than outcome. After all, Edge City is nearly off the map, a place where people go who don't fit in or measure up in conventional terms.

A Harvard grad who had taken up Zen study, James Shapiro wanted to test himself, and thus define himself, by a really hard and lonely effort. He chose running. There was no payoff, little to be gained in publicity or notoriety. Only fellow hard-core runners understood the significance of this, and even they missed it if they were overly concerned with their own race times. Shapiro started in California and ran to New York City, his hometown. Somewhere he entered

Edge City, that place of supreme effort and self-sacrifice. He learned that it is not a place but a process, being alone with himself during the many difficult days of plodding forward. He learned that he could do it, and that there is no "it," no destination—only a getting there.

This is a postmodern lesson, the idea that you are never at your destination or goal. It is also Zen. The finish line is but a subject position, not a fixed place. Elders stress goals but forget that getting there is more than half the fun; it is all there is. Kids who stay up late on the computer or cell are perhaps embarking on their own hard runs, creating themselves in the talking and posting and blogging that they are doing. Shapiro learned that he was never alone; people inhabited his run and helped him out. The kids know better than we do that the body politic is made up of other bodies, the people on the other end of text messages. They are innately social in ways that we were not, those of us who grew up in an age of rugged individualism, preceding the global village.

A sociology of effort would reveal that there is a thin boundary, so thin as to be fragile and permeable, between effort and compulsion. So-called Internet addiction (Young and de Abreu 2010) describes the lives of many young and old alike. But we need to redefine work and leisure to include living our lives on the screen, which are writing lives, not that much different from the life that Norman Mailer (2004) led when he rented a spare room in Brooklyn, a garret, in which he could write. He wrote in silence and alone, whereas our kids compose in the noisy rooms of the Internet, with iPods and laptops blaring their music. One could view the age of distraction as the age of multitasking, of juggling and balance. As with most things generational, this will largely depend on when you were born.

Slacking is the absence of effort, most of the time. Hating school is not an index of giving up, of a lack of passion. Schools are not Edge Cities; they are routinized and prosaic. Little intellectually exciting happens there as teachers are strangled by a standard curriculum and must teach to the test. Kids are just biding their time, waiting for their sentences to end. The adults in the building are prisoners, too. That is what many slacking kids don't get. The teachers and parents aren't wardens; they are equally alienated. They occupy different prisons in which the time ticks away and they imagine being out on the open road, like Shapiro. Edge City is chimerical; there is homework to be done, meals cooked, memos written, quotas met. We are all immersed in the quotidian.

In this context of confinement and linear time, the self is the biggest casualty. There is no first person, nor second or third persons. Identity is whittled down by command and control, what Foucault called discipline. When my son was in junior high and high school, he used his computer for escape and creation, playing his games, texting, messaging, writing his stories. He lost himself at school (except in tennis) and found himself at home, in late afternoon and at night. His work was social, but only when he was most alone. He connected with his friends and strangers, too, in an online writing community. School was to be endured. He is not a slacker on the outside, but on the inside he yearns for greener pastures that allow him to think, write, socialize. One day, he asked me whether tomorrow would be a snow day. I asked why. He said he needed a break from school, a sentiment with which I sympathized. As a professor, my days are varied, restful, and under my control. His school days were a regime of tasks, assignments, duties. A bell rang and he moved quickly from class to class, his identity fading until he got home to his phone, music, and computer—his canvases. Now that he is in college, he loves school; he is free, his professors treat him with respect, he is not always exhausted and overburdened.

College is not perfect, only better. Some faculty (with low self-esteem) exhibit the same authoritarianism as many teachers in the earlier grades. They pile on the homework, teach to the test, and view young people as the enemy. "Gotcha! You missed that detailed question; you were late to class; your grade will suffer." Empathy does not always abound. Caring professors are sometimes dismissed as "popular" in comparison with the hard asses who treat their students as inferiors and even as criminals. Faculty with low teaching evaluations (from students) are often heard to mutter about the importance of maintaining standards.

Since Sputnik and the more recent challenges to American capitalism posed by the Japanese and Chinese, we have allowed education to become instrumental. We worry about metrics that measure outcomes: test scores, numbers of graduates in science and engineering, gross national product. But the self is not an outcome that can be readily measured. Even our college curricula are succumbing to the measurement business; I must include measures on my syllabus that allow me to evaluate "learning outcomes." Dumbing down is everywhere.

Instead of defining education as what kids learn and how much they remember, we might better view it as generative. Generative learning is lifelong learning, and it is primarily about learning who you are meant to be. It is like a run across the country: the end is the getting there, the moments of self-discovery instead of the arriving in Central Park, crossing a finish line. There is no end, only the ways in which we inhabit the world, at home wherever we may be. A run that long cannot be swallowed in one gulp; it stretches over months, like any project worth doing. Edge Cities overlap all the other places where we dwell; on Tuesday, we may run or paint or type hard, while on Wednesday we ease up. Creation and compulsion shadow each other and vie for the ever-moving self.

Let me conclude with thoughts about what we can do to improve the lives of kids and how we can learn from kids about our better selves. We can all text toward utopia, where blogging is any self-expressive activity linking us to the world and to others. Blogging is literary activity, but it could include science, music, exercise. These activities can take us through Edge City. We are not maximizing our potential but learning about our potentials and preferences. Taking music lessons and then practicing hard for a few years may lead us to other pathways. The music is not forgotten; we remember what it means to stay seated and to focus, hour after hour. And we never lose the love of Dvorak or the Doors. But we may learn that we are meant to write or run—or both. Here are some ways to achieve it:

- We need to count kids' writing and self-expression as "real" writing. We will value or valorize this writing, not only traditional literature. Blogging, texting, and messaging can transcend sound bites and become full-fledged paragraphs and even chapters. But we must recognize that they are writing, too. And kids are writing the self.
- We need to transform schooling into a generative process in which kids find their muse and learn a broad spectrum of literacies. We should deemphasize evaluation and minimize objective testing and teaching. Ideally, we would reunite mind and body by providing time during the school day for embodiment—students and teachers all exercising together, perhaps first thing in the morning. This would reduce attentional problems, boost self-esteem, reduce depression, connect people, and improve overall wellness (addressing obesity as well).
- We need to realize that children are free. They are creative and open-minded. They learn to be narrow. They acquire prejudice. Children's freedom should not be envied and then resented and extinguished. It should be encouraged and emulated by adults who are in touch with their child-like natures. Children, when unencumbered, play. Adults also need to play, creatively and in ways that unite mind and body. Disembodiment is an enemy of health and self-esteem.
- We need to be alert to the thin boundary between creation and compulsion, especially among our children. Too much time spent with

information, communication, and entertainment technologies confines children too much (and adults, too). It is not enough to idealize the past, a golden age perhaps, when kids went out to play. That is good, but there are other ways to formulate utopia: young people playing with concepts, images, and their own bodies. This can be inside and outside activity, and adult activity, too. Break up the day with exercise and play. View work as play and play as work, especially where it takes us to Edge City.

- Value both the well-rounded and lopsided self. People need to try many things to learn the one or two things that will define them. Nothing says that we must play the violin and compose poetry, write for the school newspaper and work for Habitat for Humanity. Service is good, as are a broad portfolio of activities, but these will eventually focus into the one thing you really enjoy. And real enjoyment, as I am terming it, lies along that long road from one coast to another, passing through effort, exertion, exhaustion. You start rounded, and end up lopsided, but lopsidedness calls upon the earlier activities that you shed as you grow a self.

- Encourage kids not to form countercommunities and speak in prison code. Adults must demonstrate that they are not the enemy. Much of this involves the clash between adult Puritanism and youthful sexual experimentation. Facebook can be pornographic as children recount their urges and experiments. Much of this is wish fulfillment. They need to feel free to build adults into the conversation. We were young once, and we can empathize and not just disapprove and punish.

- Respect boundaries but make them fluid. Adults shouldn't simply permit anything and everything. Some activities are self-destructive, even suicidal. Boundaries have been a casualty of the postmodern moment but must be brought back as well as reevaluated. My wife and I want to be the pre-postmodern parents who establish boundaries but are egalitarian enough that we talk about them, defending and sometimes modifying them. It is not cool to drive drunk. Piercings and tattoos are disserving later in life. "Friends with benefits" usually means that the girls give and the boys take. Boundaries shouldn't be thoughtless or authoritarian, but they are important.

- Adults need to model all of the good things: healthfulness, freedom, self-expression, community building, play. Alienation is reproduced across generations, sometimes through sheer imitation. Kids need to watch us write poetry, go for a walk or run, join a social movement, help others.

They need to know that we were young once and were in existential crisis. That we didn't get perfect grades or test scores. That we changed our college major several times. That we sometimes dislike our boss and how we deal with that. Kids and adults may not be friends, exactly, but intergenerational solidarity takes the edge off for kids for whom childhood is a struggle at the best of times.

- We need to show kids that a healthy dose of civil disobedience can go a long way. Authoritarian teachers can ruin your child's day. Standing up to them is like facing down a bully. Kids need to know that their parents support them when they contest a grade, a dress code, adult arbitrariness. Kids, like prisoners, are cowed. They hate but do not contest adult authority. Slacking might be defined as acquiescence—to parents, teachers, one's seeming fate. Secretly, the slacker wants to rebel, even to create a revolution. But he or she does not yet possess the vocabulary with which to utter these words. Our high schools do not teach Camus, let alone Frantz Fanon. There is no pedagogy of the oppressed. Instead, we recite the pledge of allegiance.

- When kids turn sixteen, adults must not just toss them the keys to their car and wash their hands of their growing up. There is far to go. Using Shapiro's metaphor, they are only in Iowa, not yet close to the eastern seaboard. What happens between sixteen and college could be fateful: challenges met, adventures undertaken, times of great excitement and also heartbreak. Adults have their own stresses, especially in this economy and in light of broken families. Divorce makes adolescents of adults, who already have adolescents of their own. In this context, everyone needs parenting. Everyone needs to be heard.

A list of desiderata could be many pages long. Listen to kids, read their writing, avoid censorship, valorize and validate them, draw and debate boundaries, slow down growing up, acknowledge the child within.

Perhaps the family could start a family blog—a nerdy idea to be sure! Adults can have a hard time with the kind of openness that blogs entail, with potentially millions reading about us, including the embarrassing details. Are blogging, texting, and messaging narcissistic? By yesterday's Puritanical standards, certainly. I share this aversion, sometimes, when I read a cloying page written by an adult who should know better. But perhaps the blogger has solid mental health, writing his life and thus both calming down and reckoning with the tough parts.

Shapiro's run was not all Zen revelation. Much of it was hard and thought-less; that is, he had run out of things to think. The running had run him out of ideas. It does that, as one becomes one with the task, putting one foot in front of another. When I run ten or more, this happens to me, too. I just give in to it. It can also happen in writing, when the words flow and they seem not to come from conscious deliberation. And then they stop, having run their course. Per-haps those are the moments when clarity is achieved and oneness with the world attained. I know from my running that there is a certain state of flow, beyond the warm-up but before exhaustion, when I experience a oneness of mind and body, and with the world, that is beyond words. We are not all runners. There are days when the pages of the journal, blog, story, or article will be next to worthless. Or there won't be any words at all. I am convinced that writing and running are keys to my mental health. When I do neither, I am lethargic. I might even sulk. When I do both, I come alive or, to use the metaphor of the transcontinental run, I maintain. A good, easy pace that I could maintain forever, or at least until dark, when I find a place to sleep.

I loved Shapiro and Pirsig when I first read them because, for me at least, the message of the 1960s was Zen: be where, and who, you are. Don't let adults determine your path, undermining your self-esteem. What I have learned since then is that my own parents, liberal movement people, were a step ahead of me. They understood that parenting involves letting kids make mistakes, learning from them, and for this reason I turned out like them, for better or worse. I think it is for the better because young parents then were likely as not commit-ted to a vision of a better world and willing to make sacrifices to see it happen. I am sixties, through and through: I mistrust authority, don't stand on ceremony, value consensus, dislike injustice.

We baby boomers can easily forget that we were young once. Social media might seem alien and apolitical, time misspent. But our kids are doing what we did: carving out a place in a world that basically ignores them. We were busy writing our own manifestos, school newspapers, love letters. Our kids are doing the same things; they are passionate, and they put their passion into words. They might not turn in their papers on time or even write them. But their textual and posttextual work count as literary work, and their social networking is community organizing by another name.

REFERENCES

Adam, Barbara. 1998. *Timescapes of Modernity: The Environment and Invisible Hazards*. London: Routledge.

———. 1995. *Timewatch: The Social Analysis of Time*. Cambridge, UK: Polity.

Adorno, Theodor W. 1974. *Minima Moralia: Reflections from Damaged Life*. London: Verso.

———. 1973a. *Negative Dialectics*. New York: Continuum.

———. 1973b. *The Jargon of Authenticity*. Evanston, IL: Northwestern University Press.

———. 1973c. *Philosophy of Modern Music*. New York: Seabury.

Agger, Ben. 2012. *Oversharing: Presentations of Self in the Internet Age*. New York: Routledge.

———. 2010. *Body Problems: Running and Living Long in a Fast-Food Society*. New York: Routledge.

———. 2009a. *The Sixties at 40: Leaders and Activists Remember and Look Forward*. Boulder, CO: Paradigm.

———. 2009b. "Text Messages: Reading Kids' Writing Politically." *New York Journal of Sociology* 2, no. 1. http://newyorksociology.org/.

———. 2005. "Beyond Beltway and Bible Belt: Re-imagining the Democratic Party and the American Left." *Fast Capitalism* 1, no. 1. www.fastcapitalism.com.

———. 2004a. *Speeding Up Fast Capitalism: Cultures, Jobs, Families, Schools, Bodies*. Boulder, CO: Paradigm.

———. 2004b. *The Virtual Self: A Contemporary Sociology*. New York: Blackwell.

————. 2002. *Postponing the Postmodern: Sociological Practices, Selves and Theories.* Lanham, MD: Rowman & Littlefield.

————. 1990. *The Decline of Discourse: Reading, Writing and Resistance in Postmodern Capitalism.* New York: Falmer.

————. 1989a. *Fast Capitalism: A Critical Theory of Significance.* Urbana: University of Illinois Press.

————. 1989b. *Reading Science.* Dix Hills, NY: Falmer.

Agger, Ben, and Timothy W. Luke, eds. Forthcoming. *Gun Violence and Public Life.* Boulder, CO: Paradigm.

Agger, Ben, and Beth Anne Shelton. 2007. *Fast Families, Virtual Children: A Critical Sociology of Families and Schooling.* Boulder, CO: Paradigm.

Aronowitz, Stanley. 2008. *Against Schooling: For an Education That Matters.* Boulder, CO: Paradigm.

————. 1992. *False Promises: The Shaping of American Working Class Consciousness.* Durham, NC: Duke University Press.

Baudrillard, Jean. 1983. *Simulations.* New York: Semiotext(e).

Bell, Daniel. 1973. *The Coming of Post-Industrial Society: A Venture in Social Forecasting.* New York: Basic.

Bendix, Reinhard. 1958. *Work and Authority in Industry.* Berkeley: University of California Press.

Benjamin, Walter. 1999. *The Arcades Project.* Cambridge, MA: Belknap Press of Harvard University Press.

————. 1969. *Illuminations.* New York: Schocken.

Bergland, Christopher. 2007. *The Athlete's Way: Sweat and the Biology of Bliss.* New York: St. Martin's.

Bernstein, Basil. 1975. *Class, Codes, and Control: Theoretical Studies towards a Sociology of Language.* New York: Schocken.

Bogle, Kathleen A. 2008. *Hooking Up: Sex, Dating, and Relationships on Campus.* New York: New York University Press.

Borg, Gunnar. 1998. *Borg's Perceived Exertion and Pain Scales.* Champaign, IL: Human Kinetics.

Bowles, Samuel, and Herbert Gintis. 1976. *Schooling in a Capitalist America: Educational Reform and the Contradictions of Economic Life.* New York: Basic.

Braverman, Harry. 1974. *Labor and Monopoly Capital: The Degradation of Work in the Twentieth Century.* New York: Monthly Review Press.

Carek, P. J., S. E. Laibstain, and S. M. Carek. 2011. "Exercise for the Treatment of Depression and Anxiety." *International Journal of Psychiatry in Medicine* 41, no. 1 (2011): 15–28.

Carson, Rachel. 1962/1999. *Silent Spring.* London: Penguin.

Cleaver, Eldridge. 1968. *Soul on Ice.* New York: Dell.

Collins, Randall, ed. 1979. *The Credential Society: An Historical Sociology of Education and Stratification*. New York: Academic.

Coupland, Douglas. 1991. *Generation X: Tales for an Accelerated Culture*. New York: St. Martin's.

Dahl, Svend. 1958. *History of the Book*. Metuchen, NJ: Scarecrow.

Davis, Mike. 1990. *City of Quartz: Excavating the Future in Los Angeles*. London: Verso.

Debord, Guy. 1970. *Society of the Spectacle*. Detroit: Red and Black.

Derrida, Jacques. 1978. *Writing and Difference*. Chicago: University of Chicago Press.

Diamond, Stanley. 1974. *In Search of the Primitive: A Critique of Civilization*. New Brunswick, NJ: Transaction.

D'Souza, Dinesh. 2007. *The Enemy at Home: The Cultural Left and Its Responsibility for 9/11*. New York: Doubleday.

Dyer-Witheford, Nick. 1999. *Cyber-Marx: Cycles and Circuits of Struggle in High-Technology Capitalism*. Urbana: University of Illinois Press.

Ewen, Stuart. 1976. *Captains of Consciousness: Advertising and the Social Roots of the Consumer Culture*. New York: McGraw-Hill.

Flacks, Richard. 2005. "Information Technology and Participatory Democracy: Some Considerations." *Fast Capitalism* 1, no. 2. www.fastcapitalism.com.

Foucault, Michel. 1977. *Discipline and Punish: The Birth of the Prison*. New York: Vintage.

Freire, Paulo. 1970. *Pedagogy of the Oppressed*. New York: Herder and Herder.

Friedan, Betty. 1963/2001. *The Feminine Mystique*. New York: Norton.

Garfinkel, Harold. 1967. *Studies in Ethnomethodology*. Englewood Cliffs, NJ: Prentice Hall.

Giroux, Henry. 2012. *Youth in Revolt: Reclaiming a Democratic Future*. Boulder, CO: Paradigm Publishers.

———. 1988. *Schooling and the Struggle for Public Life: Critical Pedagogy in the Modern Age*. Minneapolis: University of Minnesota Press.

Gitlin, Todd. 2006a. *The Intellectuals and the Flag*. New York: Columbia University Press.

———. 2006b. "The Kids Are Alright." *Plenty Magazine*.

———. 2003a. *Letters to a Young Activist*. New York: Basic.

———. 2003b. *The Whole World Is Watching: Mass Media in the Making and Unmaking of the New Left*. Berkeley: University of California Press.

———. 1987. *The Sixties: Years of Hope, Days of Rage*. New York: Bantam.

Goffman, Erving. 1959. *The Presentation of Self in Everyday Life*. New York: Anchor.

Gould, Emily. 2010. *And the Heart Says Whatever*. New York: Free Press.

———. 2008. "Exposed." *New York Times,* May 25.

Gouldner, Alvin W. 1970. *The Coming Crisis of Western Sociology.* New York: Basic.

Griffith, Mary. 1998. *The Unschooling Handbook: How to Use the Whole World as Your Child's Classroom.* New York: Three Rivers Press.

Habermas, Jurgen. 1989. *The Structural Transformation of the Public Sphere.* Cambridge, MA: MIT Press.

———. 1987a. *The Philosophical Discourse of Modernity.* Cambridge, MA: MIT Press.

———. 1987b. *The Theory of Communicative Action.* Vol. 1. Boston: Beacon.

———. 1984. *The Theory of Communicative Action.* Vol. 2. Boston: Beacon.

———. 1971. *Knowledge and Human Interests.* Boston: Beacon.

Hall, David. 1996. *Cultures of Print: Essays in the History of the Book.* Amherst: University of Massachusetts Press.

Haraway, Donna. 1991. *Simians, Cyborgs, and Women: The Reinvention of Nature.* New York: Routledge.

Hardt, Michael, and Antonio Negri. 2005. *Multitude: War and Democracy in the Age of Empire.* New York: Penguin.

———. 2000. *Empire.* Cambridge, MA: Harvard University Press.

Harvey, David. 1989. *The Condition of Postmodernity: An Enquiry into the Origins of Cultural Change.* Oxford: Blackwell.

Hassan, Robert. 2012. *The Age of Distraction: Reading, Writing, and Politics in a High-Speed Networked Economy.* New Brunswick, NJ: Transaction.

———. 2003. *The Chronoscopic Society: Globalization, Time, and Knowledge in the Network.* New York: Peter Lang.

Hayden, Tom. 1988. *Reunion: A Memoir.* New York: Random House.

———. 1962/2005. *The Port Huron Statement.* New York. Avalon.

Heidegger, Martin. 1927/2000. *Being and Time.* Oxford: Blackwell.

Henderson, Joe. 2004. *Run Right Now.* New York: Barnes and Noble Library.

Hinduja, S., and J. W. Patchin. 2012. *School Climate 2.0: Preventing Cyberbullying and Sexting One Classroom at a Time.* Thousand Oaks, CA: Sage.

———. 2008. "Cyberbullying: An Exploratory Analysis of Factors Related to Offending and Victimization." *Deviant Behavior* 29, no. 2: 129–56.

Hochschild, Arlie. 1989. *The Second Shift: Working Parents and the Revolution at Home.* New York: Viking.

Horkheimer, Max. 1974. *Eclipse of Reason.* New York: Seabury.

Horkheimer, Max, and Theodor W. Adorno. 1972. *Dialectic of Enlightenment: Philosophical Fragments.* New York: Herder and Herder.

Husserl, Edmund. 1970. *The Crisis of European Sciences and Transcendental Phenomenology.* Evanston, IL: Northwestern University Press.

Huxley, Aldous. 1953. *Brave New World.* New York: Bantam.

Illich, Ivan. 1971. *Deschooling Society*. New York: Harper and Row.

Irwin, William, ed. 2002. *The Matrix and Philosophy: Welcome to the Desert of the Real*. Chicago: Open Court.

Jackson, George. 1994. *Soledad Brother: The Prison Letters of George Jackson*. Chicago: Chicago Review Press.

Jacobs, Gloria. 2009. *Adolescents and Instant Messaging: Literacy, Language, and Identity Development in the 21st Century*. Saarbrücken: VDM Verlag.

———. 2008a. "We Learn What We Do: Developing a Repertoire of Writing Practices in an Instant Messaging World." *Journal of Adolescent and Adult Literacy* 52, no. 3: 203–11.

———. 2008b. "Saying Something or Having Something to Say: Attention Seeking, the Breakdown of Privacy, and the Promise of Discourse in the Blogosphere." *Fast Capitalism* 4, no. 1. www.fastcapitalism.com.

Jacoby, Russell. 2005. *Perfect Imperfect: Utopian Thought for an Anti-Utopian Age*. New York: Columbia University Press.

———. 1999. *The End of Utopia: Politics and Culture in an Age of Apathy*. New York: Basic.

———. 1987. *The Last Intellectuals: American Culture in the Age of Academe*. New York: Basic.

———. 1976. "A Falling Rate of Intelligence?" *Telos* 27: 141–46.

———. 1975. *Social Amnesia: A Critique of Contemporary Psychology*. Boston: Beacon.

Kann, Mark. 2005. "From Participatory Democracy to Digital Democracy." *Fast Capitalism* 1, no. 2. www.fastcapitalism.com.

Kay, Jane Holtz. 1997. *Asphalt Nation: How the Automobile Took Over America, and How We Can Take It Back*. New York: Crown.

Keillor, Garrison. 2004. *Homegrown Democrat: A Few Plain Thoughts from the Heart of America*. New York: Viking.

Kellner, Douglas. 1995. *Media Culture: Cultural Studies, Identity and Politics between the Modern and the Postmodern*. New York: Routledge.

Kralovec, Etta, and John Buell. 2001. *The End of Homework: How Homework Disrupts Families, Overburdens Children and Limits Learning*. Boston: Beacon.

Kuhn, Thomas. 2012. *The Structure of Scientific Revolutions*. 4th ed. Chicago: University of Chicago Press.

Laing, R. D. 1967. *The Politics of Experience, and the Bird of Paradise*. Harmondsworth: Penguin.

Lasch, Christopher. 1979. *The Culture of Narcissism*. New York: Norton.

Lear, Chris. 2011. *Running with the Buffaloes*. New York: Lyons Press.

Leiss, William. 1976. *The Limits to Satisfaction: An Essay on Needs and Commodities*. Toronto: University of Toronto Press.

Leonhard, Woody. 1995. *The Underground Guide to Telecommuting: Slightly Askew Advice on Leaving the Rat Race Behind.* Reading, MA: Addison-Wesley.

Lukacs, Georg. 1971. *History and Class Consciousness.* London: Merlin.

Luke, Timothy W. 1989. *Screens of Power: Ideology, Domination and Resistance in Informational Society.* Urbana: University of Illinois Press.

Lyotard, Jean-Francois. 1984. *The Postmodern Condition: A Report on Knowledge.* Minneapolis: University of Minnesota Press.

Mailer, Norman. 2004. *The Spooky Art: Thoughts on Writing.* New York: Random House.

Malcolm X. 1999. *The Autobiography of Malcolm X.* New York: Ballantine.

Mandel, Ernest. 1978. *Late Capitalism.* London: Verso.

Marcuse, Herbert. 1972. *Counterrevolution and Revolt.* Boston: Beacon.

———. 1969. *An Essay on Liberation.* Boston: Beacon.

———. 1964. *One-Dimensional Man.* Boston: Beacon.

———. 1960. *Reason and Revolution: Hegel and the Rise of Social Theory.* Boston: Beacon.

———. 1958. *Soviet Marxism: A Critical Analysis.* London: Routledge and Kegan Paul.

———. 1955. *Eros and Civilization.* New York: Vintage.

Marx, Karl. 1867/1967. *Capital.* Vol. 1. New York: International Publishers.

Marx, Karl, and Friedrich Engels. 1848/1998. *The Communist Manifesto.* New York: Monthly Review Press.

McDougall, Christopher. 2009. *Born to Run: A Hidden Tribe, Superathletes, and the Greatest Race the World Has Never Seen.* New York: Knopf.

Merleau-Ponty, Maurice. 1962. *Phenomenology of Perception.* New York: Routledge.

Miller, James. 1987. *Democracy Is in the Streets: From Port Huron to the Weather Underground.* New York: Simon & Schuster.

Miller, Mark Crispin. 1988. *Boxed In: The Culture of TV.* Evanston, IL: Northwestern University Press.

Mills, C. Wright. 2002. *White Collar: The American Middle Classes.* New York: Oxford University Press.

———. 1959. *The Sociological Imagination.* New York: Oxford University Press.

Moreno, M. A., M. R. Parks, F. J. Zimmerman, T. E. Brito, and D. A. Christakis. 2009. "Display of Health Risk Behaviors on MySpace by Adolescents: Prevalence and Associations." *Archives of Pediatric and Adolescent Medicine* 163, no. 1 (January): 27–34.

Moreno, M. A., A. Vanderstoep, M. R. Parks, F. J. Zimmerman, A. Kurth, and D. A. Christakis. 2009. "Reducing At-Risk Adolescents' Display of Risk Behavior on a Social Networking Web Site: A Randomized Controlled Pilot

Intervention Trial." *Archives of Pediatric and Adolescent Medicine* 163, no. 1 (January): 35–41.

Murakami, Haruki. 2009. *What I Talk about When I Talk about Running*. New York: Vintage.

Negroponte, Nicholas. 1996. *Being Digital*. New York: Knopf.

Ogburn, William. 1964. *On Culture and Social Change*. Chicago: University of Chicago Press.

Oliver, J. Eric. 2006 *Fat Politics: The Real Story Behind America's Obesity Epidemic*. New York: Oxford University Press.

O'Neill, John. 1972. *Sociology as a Skin Trade: Towards a Reflexive Sociology*. New York: Harper and Row.

Paci, Enzo. 1972. *The Function of the Sciences and the Meaning of Man*. Evanston, IL: Northwestern University Press.

Packard, Vance. 1959. *The Status Seekers: An Exploration of Class Behavior in America and the Hidden Barriers That Affect You, Your Community, Your Future*. New York: David McKay.

Paine, Thomas. 1791/1985. *Rights of Man*. New York: Penguin.

Parsons, Talcott. 1951. *The Social System*. New York: Free Press.

Patchin, J. W., and S. Hinduja. 2010. "Trends in Online Social Networking: Youth Use of MySpace Over Time." *New Media & Society* 12, no. 2: 197–216.

Petrini, Carlos. 2003. *Slow Food: The Case for Taste*. New York: Columbia University Press.

Pirsig, Robert M. 1974. *Zen and the Art of Motorcycle Maintenance*. New York: Morrow.

Pollan, Michael. 2008. *In Defense of Food*. New York: Penguin.

———. 2006. *The Omnivore's Dilemma: A Natural History of Four Meals*. New York: Penguin.

Poster, Mark. 2001. *What's the Matter with the Internet?* Minneapolis: University of Minnesota Press.

Ratey, John J., and Eric Hagerman. 2008. *Spark: The Revolutionary New Science of Exercise and the Brain*. Boston: Little, Brown.

Richtel, Matt. 2010. "Your Brain on Computers: Hooked on Gadgets, and Paying a Mental Price." *New York Times*, June 7.

Rippin, Hannah. 2005. "The Mobile Phone in Everyday Life." *Fast Capitalism* 1, no. 1. www.fastcapitalism.com.

Sale, Kirkpatrick. 1973. *SDS: The Rise and Development of the Students for a Democratic Society*. New York: Random House.

Sartre, Jean-Paul. 1976. *Critique of Dialectical Reason*. London: NLB.

———. 1956. *Being and Nothingness: A Phenomenological Essay on Ontology*. New York: Washington Square Press.

Schlosser, Eric. 2001. *Fast Food Nation: The Dark Side of the All-American Meal.* New York: Houghton Mifflin.

Schumacher, E. F. 1973. *Small Is Beautiful: Economics as if People Mattered.* New York: Harper Colophon.

Shapiro, James. 1982. *Meditations from the Breakdown Lane: Running across America.* New York: Random House.

Sheehan, George. 1978. *Running and Being.* New York: Simon & Schuster.

Smith, Adam. 1776/2003. *The Wealth of Nations.* New York: Bantam Classic.

Soja, Ed. 1989. *Postmodern Geographies: The Reassertion of Space in Critical Social Theory.* London: Verso.

Szasz, Thomas. 1973. *The Age Madness.* Garden City, NY: Anchor.

Taylor, Frederick. 1947. *The Principles of Scientific Management.* New York: Norton.

Turkle, Sherry. 2011. *Alone Together: Why We Expect More from Technology and Less from Each Other.* New York: Basic.

———. 1995. *Life on the Screen: Identity in the Age of the Internet.* New York: Simon & Schuster.

Varon, Jeremy. 2004. *Bringing the War Home: The Weather Underground, the Red Army Faction and Revolutionary Violence in the Sixties and Seventies.* Berkeley: University of California Press.

Vitzthum, Virginia. 2007. *I Love You, Let's Meet: Adventures in Online Dating.* Boston: Little, Brown.

Wells, Tom. 1994. *The War Within: America's Battle over Vietnam.* Berkeley: University of California Press.

Whyte, William. 1956. *The Organization Man.* New York: Simon & Schuster.

Wittgenstein, Ludwig. 1958. *Philosophical Investigations.* Oxford: Blackwell.

Wood, Daniel B. 2013. "Powering Down: Summer Vacation and the Pursuit of Doing ... Nothing." *Christian Science Monitor Magazine* (May 27).

Young, Kimberly S., and Cristiano Nabuco de Abreu. 2010. *Internet Addiction: A Handbook and Guide for Evaluation and Treatment.* New York: Wiley.

INDEX

ABOUT THE AUTHOR

Ben Agger is Professor of Sociology at the University of Texas, Arlington, where he directs the Center for Theory. His most recent book is *Oversharing: Presentations of Self in the Internet Age.*

Several chapters have been adapted from the following publications, which have been used with permission:

Ch. 1: "The Book Unbound: Reconsidering One-Dimensionality in the Internet Age." In *Putting Knowledge to Work and Letting Information Play*, 2nd ed., edited by Timothy W. Luke and Jeremy Hunsinger. Rotterdam, NY: Sense Publishers, 2012.
Ch. 4: "The Pulpless Generation: Why Young People Don't Protest the Iraq War (or Anything Else), and Why It's Not Entirely Their Fault." *Cultural Studies <=> Critical Methodologies* 9, no. 1 (February 2009): 41–51.
Ch. 6: "iTime: Labor and Life in a Smartphone Era." *Time & Society* 20, no. 1 (2011): 119–36.
Ch. 11: "Text Messages: Reading Kids' Writing Politically." *New York Journal of Sociology* 2, no. 1 (2009).
Ch. 12: "Bodies of Knowledge: Considerations of Science, Exercise, Food and Body Politics." *disclosure* 19 (2010): 1–6.